THE
FLYING
ZOO

BIRDS,

PARASITES,

AND THE

WORLD THEY

SHARE

THE FLYING ZOO

MICHAEL STOCK

UNIVERSITY *of* **ALBERTA** PRESS

Published by

University of Alberta Press
Ring House 2
Edmonton, Alberta, Canada   T6G 2E1
www.uap.ualberta.ca

Library and Archives Canada Cataloguing in Publication

Title: The flying zoo : birds, parasites, and the world they share / Michael Stock.
Names: Stock, Terrance Michael, author.
Description: Includes bibliographical references and index.
Identifiers: Canadiana (print) 20190112905 | Canadiana (ebook) 20190114134 | ISBN 9781772123746
    (softcover) | ISBN 9781772124408 (PDF)
Subjects: LCSH: Birds—Ecology. | LCSH: Birds—Parasites.
Classification: LCC QL698.95 .S76 2019 | DDC 598.17—dc23

First edition, first printing, 2019.
First printed and bound in Canada by Friesens, Altona, Manitoba.
Editing and proofreading by Julie Sedivy.
Indexing by Susan Grant.

University of Alberta Press is committed to protecting our natural environment. As part of our efforts,
this book is printed on Enviro Paper: it contains 100% post-consumer recycled fibres and is acid- and
chlorine-free.

University of Alberta Press gratefully acknowledges the support received for its publishing program
from the Government of Canada, the Canada Council for the Arts, and the Government of Alberta
through the Alberta Media Fund.

Canada Council
for the Arts

Conseil des Arts
du Canada

Alberta
Government

Dedicated to the many people who are passionate about, fascinated by, and who understand the incredible value of the natural world, including birds and all forms of life—even the creepiest!

CONTENTS

PREFACE

As a young undergraduate student, I was studying in the stacks one day, deep in the bowels of the library at Western University in London, Ontario. Bored, I looked up and saw a book with a very strange title—*Fleas, Flukes and Cuckoos* by Miriam Rothschild and Theresa Clay. I picked it up and started reading, and I was hooked. I had not realized there was a whole world of biology out there that I had never even heard of, and that people were studying these gross but fascinating creatures.

The book made me interested enough to take an ecology course called "Parasitology" taught by Dr. Ken Bourns. Dr. Bourns was an engaging teacher—he was the first person who showed me parts of a live organism (the flame cells of a trematode cercarium) under a microscope, and he regaled his students with stories of parasites from tropical countries that caused horrible diseases in animals and humans.

With Bourns's encouragement, I did a fourth-year independent-study project on circadian migration of rat tapeworms and presented the results at a meeting of the Canadian Society of Zoology in Guelph, Ontario. Looking back now, my project was pretty weak, but I nevertheless had the audacity to criticize some work done by one of the most famous parasitologists in the world, Dr. John Holmes. Scared out of my mind at the meeting, I remember an audience member standing and lobbing a couple of easy questions my way. After answering, I felt like I was the best thing that had come along in Canadian science for a long time. Later, I was introduced to the audience member who had asked the questions—you guessed it, Dr. Holmes. It would have been so easy for John to destroy me publicly, but he kindly realized that, while I was young and naïve, I was also very excited about parasites. He let me off the hook and actually encouraged me to go to grad school.

I applied and was accepted at the University of Alberta in Edmonton. I had never been west of Windsor, Ontario, so this was a real adventure. I wanted to be a student of Dr. Holmes, but he was going on sabbatical leave for a year to study sea snakes in Australia. He suggested I apply to do a master's degree with one of his colleagues, Dr. Bill Samuel. I asked around about Bill Samuel, and I was told that he was an expert on wildlife biology and parasites and diseases of wildlife, and that he had many different research projects going on—everything from sarcoptic mites on wolves to helminth worms in skunks, and that he was very demanding and could be pretty tough on his grad students. Nevertheless, I decided that I wanted to see western Canada, work with wildlife, and learn from some of the best biologists in the world, so out west I went.

What I discovered when I got to Edmonton was that Dr. Samuel was an extremely active researcher with many irons in the fire. He did expect his students to work hard but he was also honest and straightforward. You always knew where you stood with Bill, and he did everything he could to get his students to succeed. I also discovered there was an incredible group of graduate students working with Dr. Holmes and Dr. Samuel as well as with Dr. Jerome Mahrt (a protozoologist), all housed on the ninth floor of the Biological Sciences building.

There I met a brilliant Australian student, Russ Hobbs, who was working on parasites of pikas in the mountains of the Yukon. I also met a student from the United States, Al Bush, working on the ecology of parasites of lesser scaup, and a fellow Canadian, Al Shostak, working on nematodes in deer. All of these fine scientists went on to make important scientific contributions, and I learned a lot about biology and a lot about life from these men. Other grad students, many of whom were not in parasitology but in diverse areas such as neurobiology and ecology, were housed near my carrel, and they opened my eyes to these parts of biology, of which I knew little. I will be forever grateful to them.

I decided to write this book as a promise I made to myself some time ago. I hope that readers will be left with some intriguing questions and perhaps with a little respect for some of nature's most bizarre creatures. For birders, after reading this book I think you will never look at a bird the same way again.

During my own education, I felt it would be helpful to have a book that provided students with ideas for research projects in a way that was understandable to the general public. I hope this books provides such a resource.

Also, during my PHD research with Dr. Holmes, I had to kill many birds. I have always felt guilty that the new scientific knowledge gained from that work did not adequately pay for the lives of the birds that I had killed— perhaps this book will help in a small way to repay that debt.

My colleagues at MacEwan University, particularly Dennis Pfeffer and Mark Degner, read parts of this manuscript and provided lots of advice. In addition, the excellent professionals in our library at MacEwan University (particularly Nick Ursulak and Michelle Bezenar) helped with endless interlibrary loan requests. Our science research librarian, Karen Hering, encouraged the completion of this work. I owe her my sincerest thanks. Two anonymous reviewers, conscripted by Linda Cameron and Peter Midgley at University of Alberta Press, spent long hours reviewing the manuscript and made many constructive critical comments. Finally, I owe a great thank you to Julie Sedivy, who expertly edited this book, pointed out many ways to improve its readability, and was instrumental in keeping the story lucid.

Throughout this project, my wife Debbie and daughter Heather have encouraged and supported my work. When I doubted the value of this book, they cheered me on, gave me good advice, and kept my spirits up.

1

A WORLD
ON A BIRD

A BIRD IS AN ECOSYSTEM

There is a tale told about a famous zoologist who once stepped outside, only to have his hat bombarded by the droppings of a pigeon. Unperturbed, he pointed out the offending bird to a group of his students and excitedly exclaimed, "Look—a flying zoological garden!" Regardless of the accuracy of the story, it illustrates a unique way to look at a bird.

Birds, and for that matter any other animals, do not exist as individual organisms. They are ecosystems (or zoological collections) of specialized organisms making their homes on the inside or outside of another specialized organism—in this case, the bird. The passengers are most often referred to as parasites—a term derived from Greek (*para*—"beside"; *sitos*—"food"), denoting someone feeding at the same table; the bird, in this case, would be congenially referred to as the host.

To most biologists, animal parasites are simply creatures who make use of another live animal for their nourishment and habitat. It was the ancient Romans who gave the term a more sinister twist, invoking the typical mental picture that most of us have of parasites today. Despite regular baths, good sanitation systems, and water supplied by aqueducts, Romans were rife with lice, fleas, and bed bugs. They suffered from infections of large round-worms and constantly passed sections of fish tapeworms they acquired from their diet of raw and fermented fish. Today, most of us who have the privilege to live in developed countries of the world seldom encounter parasites, but we easily imagine them as evil, horrid, slime-covered monsters that pop out of live bodies, as depicted in the movie *Alien*. In truth, parasites have such a specialized and intimate relationship with their environment (the host) that, however you may feel about lice, ticks, worms, and other such creatures, they invite you to more deeply understand and appreciate animals that perhaps you already find fascinating or beautiful—namely, birds.

It is ultimately not in parasites' best interest to confirm our worst impressions of them. For example, you might think that parasites prey mainly on sick and dying birds, but in nature it is often the healthiest and best-fed birds that host the most parasites. This paradox is explained if we remember that a parasite's very existence depends on a healthy environment. A parasite's job, in or on a host, is to survive long enough to mate and produce the next generation. This job is generally best accomplished by keeping a clean and happy household, not by wantonly and wastefully over-exploiting the environment. Over-exploitation by parasites can result in pathology and disease in the host. The host may also suffer poor body condition, changes to normal behaviour, and failure to reproduce as a result of infestation by parasites. In the long run, parasites using this strategy would be stealing resources from their own offspring, completely going against the tenets of natural selection.

From the bird's perspective, the situation may appear straightforward. Because parasites (at the very least) steal nutritional resources, natural selection should lead to a battery of defenses that evolved to eliminate parasites. Nature, however, never seems to be this straightforward. For example, the many species of bacteria and other micro-organisms that live in the

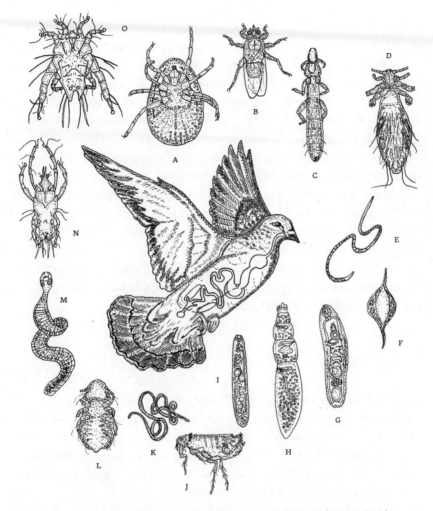

FIGURE 1.1   Animal parasites that may be found on and inside a pigeon (*Columba livia*). (A) *Argas reflexus*, a tick on the body. (B) *Pseudolynchia maura* (= *canariensis*), a louse fly on the body. (C) *Columbicola columbae*, a louse on the contour feathers. (D) *Hohorstiella lata*, a louse on the head. (E) *Capillaria columbae*, a roundworm in the crop. (F) *Tetrameres fissipina*, a roundworm in the proventriculus. (G) *Brachylaima mazantii*, a fluke in the duodenum. (H) *Davainea proglottina*, a tapeworm in the duodenum. (I) *Tanasia bragai*, a fluke in the kidneys. (J) *Ceratophyllus columbae*, a flea on the body. (K) *Splendidofilaria columbensis*, a roundworm in the thighs. (L) *Campanulotes bidentatus*, a louse on the downy body feathers. (M) *Raillietinia micracantha*, a tapeworm in the duodenum. (N) *Diplaegidia columbae*, a mite on the contour feathers. (O) *Falculifer rostratus*, a mite on the contour feathers.

digestive systems of animals (including ourselves) may provide a number of benefits to their hosts, such as protecting them from being infected by much nastier microbes, providing a source of vitamin K, and stimulating and tuning the host's immune system. Similarly, parasitic animals may provide services to their hosts—there is evidence that infection by roundworms and tapeworms can prevent or alleviate autoimmune conditions and allergies in the host. Even from a purely energetic viewpoint, perhaps the loss of small amounts of energy to host a few parasites may be greatly outweighed by the caloric costs incurred in trying to eliminate them.

My intention in this book is to explore these complex and often contradictory notions. How has this weird association between one organism (a bird) and its fellow travellers (parasites) become normal? What special adaptations have parasites had to evolve to be able to find, colonize, and survive in or on their hosts? What effects do these parasites have on birds? How do parasites get along together in the "zoo"? How have hosts evolved to survive with their "zoological garden"?

A PARASITE'S PARADISE

A host bird that is a Garden of Eden to one parasite may be a hostile environment to another. Hence, the properties of a parasite as well as its place of residence on a bird's body often result from mutual adaptations between the partners: the bird attempts to rid itself of noxious parasites, and the parasite changes its form and behaviour to evade these attempts. These interactions have a profound effect on bird and parasite alike.

One set of adaptations revolves around the efforts that birds make to preserve some of their most important assets: their feathers. On a bird, different kinds of feathers are distributed in different body regions and have different functions. Contour feathers, with a shaft and vane, are used for flight—those on wings are called remiges and those on tails are retrices. Other feathers called semiplumes (with a shaft but with downy barbs) are interspersed with contour feathers. Insulating down feathers, with barbs occurring as fluffy tufts, occur all over the bodies of hatchlings and as warm undercoats on adults. Sensory filoplumes are hairlike and occur among contour feathers. Rictal bristles, which are also hairlike, occur around the base of the bill (called the rictus). Consequently, parasite-induced damage to feathers can interfere with flight, cause increases in metabolism to compensate for heat loss, and even affect behaviour.

Tail feathers allow birds to balance, steer, and brake during flight. In addition, they are often important for territorial and courtship displays. The famous fossil "bird," *Archaeopteryx lithographica*, which had characteristics of both dinosaurs and birds, had tail feathers. Today, some birds, such as grebes (from the family Podicipedidae), have reduced tails and are clumsy flyers, while others, like the tropical quetzal (*Pharomachrus mocinno*), have magnificent trains and can swoop and turn and perform amazing aerial acrobatics during courtship.

Feathers are of such critical importance to birds that they must spend lots of time caring for them. Preening removes and renovates damaged feathers. Some birds, such as grebes, eat their preened feathers, possibly to create a digestive block that helps them to digest coarse food like fish bones. Preening keeps feathers clean and in order, and ensures that the barbs of contour feathers are locked together to make a good airfoil. Preening can also spread waterproofing oils on the feathers so that birds don't get soaked. Soaked feathers will not allow a layer of insulating air to be trapped between the skin's surface and the feathers. With wet plumage, birds can rapidly lose heat energy and suffer from hypothermia.

This all-important preening behaviour can also be used to reduce a bird's parasite population. In fact, birds' bodies have been tweaked by evolution to make preening especially effective for this purpose. For example, the degree of curved overhang at the tip of the upper part of a bird's bill has probably evolved mainly for the purpose of removing feather lice.[1] Less directly, the secretions of the preen gland that a bird spreads on its feathers for waterproofing also inhibit lice and other insects due to its repellant chemicals.

Tail feathers are often important as a sexual display during courtship and can signal valuable information about a bird's health and hygienic habits. Females might select males that have the longest and showiest plumage because they think these males are least affected by parasites, are

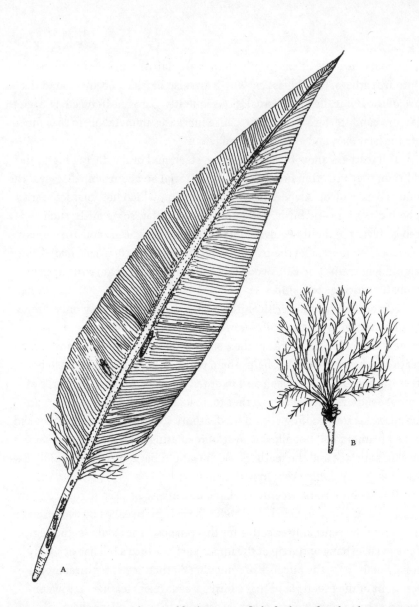

FIGURE 1.2   The structure and types of feathers. (A) A flight feather infested with ectoparasites, which live on the outside of the host. A feather louse lies parallel to the barbs of the leading edge (anterior vane) and another is nestled in the groove between the rachis (shaft) and the barbs of the posterior vane. Three quill mites live inside the quill (calamus) below a small opening, called the superior umbilicus. (B) A down feather. At the base of the feather are several louse eggs (nits).

FIGURE 1.3 The elaborate tail of the quetzal (*Pharomachrus mocinno*), compared with the tail of a pied-billed grebe (family Podicipedidae, *Podilymbus podiceps*), with a chick nestled in its plumage.

generally healthiest, and therefore most likely to sire offspring with the best genes for resisting diseases.

The idea that tail feathers can provide information about the contents of a male's "zoo" is supported by a study by A.P. Moller examining barn swallows (*Hirundo rustica*) and two types of parasites that infest them, mites and lice.[2] Moller showed that female swallows preferred to mate with males with longer tails; male swallows with longer tails turned out to have fewer parasites. In another study, Mati Kose and colleagues found that white spots on the tail feathers of male swallows (a place where lice preferred to feed and make holes) acted as reliable signals to females about male quality.[3] Therefore, lice (and other parasites) may affect the reproductive success of their hosts.

Charles Darwin thought that the elaborate tail ornaments and brightly coloured feathers of birds probably exposed their owners to danger from predators and were expensive to grow and maintain. They also require a lot of time to care for,[4] so a female bird that sees a male with a beautiful, showy tail sees a potential mate that is giving an honest signal that he will be a good provider and has a healthy relationship with his parasites.

To a parasite, its host's preoccupation with preening presents a grave threat. This is especially true for chewing lice, which cannot fly. Lice spend their entire lives (from egg to adult) within the feathery plumage of their host and have become specialized as a result—an analogy would be a mammal or bird that has colonized an island. Just as there are specialized species of finches, mockingbirds, and marine iguanas endemic to the Galapagos Islands, feather lice have become endemic to individual types of host birds.

Many of the physical features of lice are related to their host's characteristics and to the selection pressure of preening. Larger hosts tend to have larger lice; for example hummingbirds are infested by the *Trochiloecetes* species, some of which have a length of just 1.6 mm, while eagles are infested with the *Laemobothrion* species, which may be more than 20 mm long. This pattern is true particularly for lice living on the tails and wings of birds and is related to the hiding places that are available on birds of different sizes.

The colours of feather lice are also influenced by their habitat. White lice are found on the white plumage of swans and gulls, black lice live on coots and ravens, and yellow lice inhabit some orioles—probably by acquiring carotenoid pigments from feathers they eat. Colour matching provides camouflage for lice, helping them hide from a bird's sight as it preens. Lice that live on the bird's head, however, are not camouflaged—they can't been seen by the host.[5]

Lice also adapt to their place of residence on the bird. As noted by Theresa Clay, feather lice tend to fall into one of two extremes in body structure, depending on where on the bird they live.[6] Their body types are likely related to how easily these locations can be preened. The wings and tails of most birds are readily accessible to preening. Thus, lice that occur there have long, flat bodies with narrow heads, long legs, and weak claws. These lice can move fast and are very agile—they escape preening by getting out of the way. However, the heads and necks of birds are much more difficult to preen. There, lice have short, stout bodies, broad heads, short legs, and well-developed claws. They move slowly and try to escape preening by hunkering down and weathering the storm.

The behaviour of lice, as well as their form, is affected by the host's preening behaviour. Lice tend to move away from light and toward heat, prefer their bodies to be touching something, and have chemosensors to smell their hosts. Consequently, they nestle into the plumage and seldom wander off their bird. As many duck and grouse hunters have observed, after a bird has died, the change in the host's temperature (and likely its smell) causes lice, especially those with the long, narrow body type, to roam over the surface of the feathers. These homeless lice are desperate to find another warm body and are not averse to climbing onto a hunter's arm.

Other adaptations reflect the diet that is available in a parasite's habitat. For example, feather lice feed on pieces of feather that they clip using unspecialized mouthparts. Besides feathers, lice may also feed on dead cells and other debris at the skin surface, and some have taken to feeding on blood. Lice seem to prefer to feed on down or the downy part of larger feathers and on newly emerging feathers. During feeding, they hang on to a feather with their second and third pairs of legs, while the first pair is used to manipulate food.

Feathers are composed mainly of the protein keratin, a substance that is very difficult to break down. The enzymes necessary for its digestion are

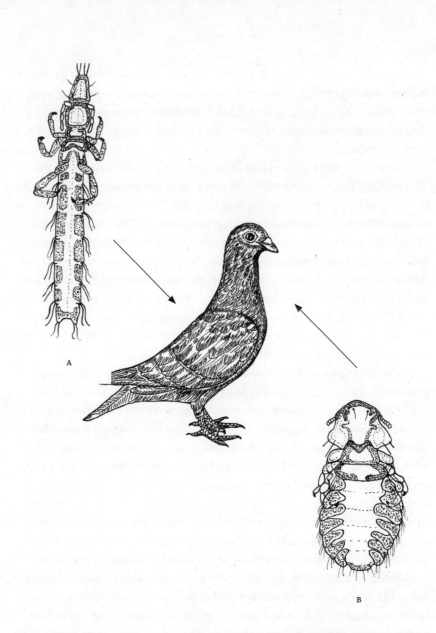

FIGURE 1.4 Two types of feather lice (Phthiraptera) found in the feathers of pigeons, showing how the body shapes of lice are adapted for their habitat on the bird in a pattern that is also seen in other birds. (A) The longer, narrower, and faster louse (*Columbicola*) is found on the wings and tail. (B) The shorter, rounder, and slower louse (*Physconelloides*) is found on the head and breast.

most likely not produced by lice, but by bacteria—called endosymbiotic bacteria (*endo*—"inside"; *symbio*—"living together")—that are housed within special cells in lice. Therefore, lice are themselves hosts of other organisms, though the dynamics of this interaction are complex and mysterious. Bacteria inside female lice live within the reproductive organs and get inside lice eggs (called nits), so that young feather lice are already armed with their complement of keratin-digesting bacteria when the nits hatch.

Trying to reproduce on a flying, preening animal poses its own challenges. Lice have what are called direct life cycles, needing only a single host to produce another generation. Adult males and females must make contact in the dense plumage to mate. Females deposit nits between barbs or at the base of feathers and cement them in place. The eggs must be securely fastened to avoid being dislodged during flight or damaged by a bird's bill during preening.

Lice make up only a small part of the flying zoo. Many other parasites live on or in birds, each with its own specific characteristics and unique relationship to the host bird. But even this small glimpse of the zoo gives us an idea of how parasites have had to evolve many special adaptations to live on a bird—and how birds in turn have had to evolve features and behaviours to live with them. An awareness of these intricate relationships allows us to see birds with completely new eyes.

THE FLYING ZOO IN MY BACKYARD

From the La-z-Boy chair in our family room, I look out into my small urban backyard, and I have a lot of fun watching the comings and goings of local birds at my feeders. One day, I had no fewer than fifteen different species of avian visitors—and this is inside a large city! Dramas play out at the feeders all the time. City-wise and street-tough house sparrows crowd out more retiring birds, like the slate-coloured juncos and red-breasted nuthatches, and then wastefully strew seeds from the feeders onto my lawn. There, shyer birds such as chipping sparrows and red-winged blackbirds get a share of the booty that has showered down from above. In the brief moments when all is quiet, beautiful house finches with their bright red heads, goldfinches, and robins sneak in for quick snacks. Noisy blue jays show up in small flocks and have a distinct pecking order at the dried corn cobs I put out for them. I am always surprised that they peck off the corn kernels in perfect rows rather than haphazardly—I wonder why? If you watch the jays carefully you can identify individuals and find out who's the boss and feeds first. Small, busy flocks of black-capped chickadees dart in and out of the feeders, constantly chirping to each other. Like the jays, I know they take some seeds and hide them for winter—the chickadees in northern areas with severe winters have especially big brains (and small digestive tracts) so they can remember where they cached their food.[7] Of all the birds at my feeders, my favourites are the blue jays and black-capped chickadees.

As I look out at this scene, I am captivated not just by what I can see but also by what I can't see. My work as a scientist who studies bird parasites causes me to wonder about the hidden part of the drama unfolding before my eyes: the flying zoo that makes each bird what it is. As I gaze out at my favourite birds, I wonder what role their parasites have played in shaping their fascinating behaviours and alluring appearance.

Could the vivid blueness of blue jays somehow be related to the flying zoo they house? Both male and female blue jays (*Cyanocitta cristata*) have gorgeous azure crested caps, with white cheeks outlined by jet black lines, one of which passes through the eye. (These black marks help you recognize individuals—but look carefully, as the marks vary between the right and left sides of their heads.) The napes and backs of blue jays are also azure, but their wings and tails have patches of sky blue and white, defined by black bars. Their breasts and bellies are buff and creamy white.

Blue colours on birds are not due to chemical pigments—they result from structural effects.[8] As feathers grow, each cell produces stringy keratin proteins. The keratin separates from water inside the cells, and after each cell dies, the water dries and is replaced by air. As a result, each cell is like a little compartment with strings of keratin and pockets of air. When light strikes the feather, the amounts and arrangements of keratin strings and air pockets cause red and yellow light (the longer wavelengths) to cancel each other, while the shorter blue wavelengths are amplified and reflected. This creates different shades of blue and sometimes iridescence and reflection of ultraviolet (uv) light (which we cannot see, but birds can).

Studies of other blue birds offer some clues about the information that is conveyed by their blue colour and its connection to the birds' parasites. For example, Australian satin bowerbirds (*Ptilonorhynchus violaceus*) are sexually dimorphic—males have metallic, shiny, blue plumage, and violet-blue eyes, while females are greenish-brown, also with blue eyes. Along with other parasites, satin bowerbirds are infested by feather lice (*Myrsidea ptilonorhynchi*) on their heads near the eyes, where these lice easily escape preening. The lice eat feathers in addition to feeding on the bird's blood and skin. They are passed from bird to bird during body contact. Satin bower-birds are also infected by blood parasites, *Haemoproteus*, which make the birds sick and reduce their stamina.

Naturally, female satin bowerbirds would prefer a mate that is not riddled with lice or blood parasites. The colour of their prospective mates—along with their elaborate courtship behaviours—may help them make their choice. On the forest floor, male birds clear display courts and build elaborate bowers from sticks. To attract and court females, they decorate the bowers with colourful objects such as fruits, flower petals, pebbles, and human debris. The quality of the bower (in terms of stick size, density, sym-metry, decorations, and quality of construction) is related to the blue uv plumage of the male that built it—the better the bower, the more intense the plumage.[9] The quality of the bower and the plumage of males (especially on their rumps and tails) are also significantly related to infection levels of lice and blood parasites, so females can assess the health and physical condition of males using all this information. For satin bowerbirds, a more intense blue means that a male is healthier and will be more likely to be selected by a choosy female as a mating partner.

Another blue bird that has been studied is the European blue tit (*Cyanistes caeruleus*). Both sexes of these beautiful little passerines have azure blue crowns and dark blue lines that pass through the eyes and encir-cle their white cheeks. They have white foreheads and blue napes, wings, rumps, and tails. Their backs are yellow-green, and the underparts are sulfur yellow with a dark line through the middle of their abdomens.

To us, males and females look alike, but, like bowerbirds and blue jays, the blue reflects UV light, so to the eyes of other blue tits, males have a much brighter crown than females. There are a lot of colours and patterns on blue tits, and it turns out that these convey different kinds of useful information. Birds with duller white cheeks were found to be more heavily infected by a particular blood parasite (*Plasmodium*) than birds with brighter cheeks.[10] Birds with duller blue tail feathers were more heavily infected by another blood parasite, *Leucocytozoon*. The more access to resources a bird had, the brighter their yellow breast feathers. The UV reflectance of blue feathers was also related to infestations of feather mites (*Proctophyllodes stylifer*), which occur on the wing and tail feathers. Because both sexes of blue tits have these colour patterns, there is a lot of information being passed among birds with regard to their reproductive condition, fitness, dominance, and health.

What about the blue jays at my backyard feeder? Is anything known about the connection between their blueness and their resident parasites? Although blue jays are infested by one flea, six species of lice, five types of ticks, and eight species of mites, in addition to being infected by nine kinds of flukes (trematodes), three tapeworms, one acanthocephalan (thorny-headed worm), and sixteen kinds of roundworms,[11] no one has looked at blue jays to see if their plumage is signalling something about their health or parasite loads, or how their plumage affects their attractiveness to the opposite sex.

A little bit is known about Steller's jays (*Cyanocitta stelleri*), birds that are closely related to blue jays. Jeffrey Zirpoli and his colleagues have studied these birds to see whether their non-iridescent plumage colour, the width of their blue-black wing bars, and the size of their crest is related to their condition (as measured by daily rates of feather growth) and to diseases caused by various species of scaly leg mites, *Knemidokoptes*.[12] (These mange mites can cause leg and toe inflammation, toe malformations, and bleeding wounds that result in reduced feeding and poor health and condition in jays). Male jays that were in the best condition during moult had the bluest feathers, but their plumage was not found to be directly connected to parasite load. This could be because parasite load by itself is not the most accurate indicator of the birds' fitness. Zirpoli and colleagues argued that male Steller's jays that produce high levels of testosterone may have reduced immune function, which results in more severe mange even if they do not house especially large populations of parasites. No testing was done to see if mangy birds were less attractive as mates.

These studies are just enough to arouse my curiosity about what other blue jays see when they consider the plumage of their fellow jays crowded around the feeder. I will have to wait for more research to fully satisfy that curiosity.

My other favourite feeder bird, the black-capped chickadee (*Poecile atricapillus*), is not very colourful, despite being related to the blue tit. Males and females have similar markings—black bibs and caps with white cheeks and underparts. Their flanks have a buffy, brownish-red tinge, and their tails are dark gray, flanked by creamy white margins. They are small—only 12 cm to 15 cm in length, and 9 g to 14 g in weight. Besides the iconic "chick-a-dee" call, which they use to signal danger, to mob an enemy, or just to stay in touch with each other, they sing short melodious songs (a former student of mine called them "cheese-bur-ger" songs) to attract mates and stake out territories. It doesn't take much watching before you notice that some chickadees can displace others from the feeders, and that these birds usually have brighter white on their cheeks and larger, blacker bibs than the birds they shove aside. Are these birds also assessing each other based on their plumage? What do their simple black and white colours mean?

At first glance, male and female chickadees appear to be similar, but in reality they are sexually dimorphic.[13] Males have brighter white plumage and larger black bibs, and the contrast between these areas is greater in males than in females. The rank of a male on the chickadee social hierarchy is related to the reflectance of its white plumage—brighter males rank higher. These males are preferred by females both as steady mates and as on-the-side sexual partners. However, in addition to signalling social status, their plumage provides a health report.

For such a cosmopolitan bird, it is surprising that so few studies have investigated the black-capped chickadee's parasites and diseases. There are anecdotal accounts of blowfly larvae on chicks in nests, a report of a feather louse (*Ricinus*) that usually occurs on corvids (I wonder if a chickadee picked one up while mobbing), and reports of nest parasites typically found on brown-headed cowbirds, four kinds of generalist fleas (found on many types of birds), and feather mites. Blood parasites, however, are better known. In Alaska, 24% of chickadees were found to be infected with malaria caused by *Plasmodium*. Bird malaria is known to affect reproductive success and mate choice in other birds. One interesting finding from Alaskan chickadees was that malaria was associated with another problem—avian keratin disorder.[14]

Avian keratin disorder was first seen in Alaska in 2010, but it is now known to affect birds (chickadees and others) throughout the entire Pacific northwest, and it seems to be rapidly spreading—there has been an uncon-firmed report from southern Alberta. The disorder causes uncontrolled production of keratin and results in massively elongated and grotesquely twisted beaks and in lesions on legs and feet. No one knows the cause of the disorder, but there is speculation that it results from a viral infection. One thing that is known, however, is that chickadees with avian keratin disorder in Alaska were more than two-and-a-half times as likely to be infected by bird malaria as those without the disease.[15] Due to its effects on beaks, this keratin disorder affects preening and foraging, and it may depress normal chickadee immune function, making the birds less able to control malaria. Also, birds with deformed beaks have dull, dirty feathers and lots of feather mites.[16] The dirty feathers affect uv reflectance and decrease colour expression—which in turn affects social dominance and mate choice.[17]

When I look out my window, it shocks me that even for such common and widespread birds as blue jays and chickadees, we know so little about their health, diseases, and parasites. I can see mating and social dominance displays, and I marvel at ornaments like blue jay crests and chickadee bibs, at the colours, plumage patterns, and behaviours of my backyard birds—and then realize that all these are affected by parasites. These features are all sending honest messages to other birds about their fitness and condition and about the kind of mates and parents they will be. I don't have to go to some exotic location like the Amazon rainforest or the Galapagos Islands to see this happening—it's all happening in my own backyard. The ecological tapestry of bird behaviour and structures has been moulded, at least in part, by the intri-cate co-evolution of parasites and their host birds—by the flying zoo.

LICE

It's a Beautiful Life

AN INTIMATE CONNECTION

Lice are beautiful! To zoologists (and other scientists), beauty in nature isn't merely a matter of brilliant colours and interesting shapes—it is best exhibited by patterns. Feather lice play a starring role in an intriguing array of ecological and evolutionary patterns, despite being rather unassuming animal parasites that quietly spend their lives nestled in a bird's plumage. These patterns hint at an intimate evolutionary bond with their host birds. For example, chances are good that the ancestors of a modern louse made their homes on the ancestors of that louse's current host. Moreover, recent species of birds tend to be infested by recent species of lice, whereas more primitive species of birds have more primitive lice. And lice and their hosts are also aligned in their physical characteristics; relatively large birds tend to have relatively large lice, and lice often match their hosts in colour. As we will see, all of these patterns can be explained by the particular lifestyles and habits of lice, but they ultimately spring from the general principle that explains most observations in biology: evolution by natural selection.

Lice are insects that belong to the order Phthiraptera. They have no wings and, although they have several distinct stages of development, they do not undergo the complete metamorphosis of insects such as butterflies. The best guess is that they evolved from ancestors similar to booklice (of the order Psocoptera), which are found living in and under bark as well as in moist burrows and nests that contain the algae and fungi consumed by these insects.[1] When books are stored in damp places, booklice feed on the moulds that develop in the bindings. Although they are called "lice," these psocids are not parasites, though they are sometimes found hitching rides on birds. Because these small insects live on bark or plant foliage where birds are regularly found and have generalized mouthparts with mandibles that allow them to hitchhike on birds, they are poised to evolve into full-fledged parasites.

Feather lice apparently made the leap to full parasite status long ago in their evolutionary history, as shown by a 44-million-year-old fossil of a remarkably well-preserved specimen.[2] In fact, David Grimaldi and Michael Engel believe that tropical barklice and parasitic lice diverged at least 100 million years ago, probably even as early as the beginning of the Cretaceous, 145 million years ago.[3] Given the recent discoveries of feathered dinosaurs, it is only a matter of time before the first verified specimen of a dinosaur louse is found. (In fact, excitement was stirred up in 1999 when Alexander Rasnitsyn and Vladimir Zherikhin found a giant, early-Cretaceous louse-like fossil named *Saurodectes* in Russia, and suggested that it may have fed on pterosaurs or infested the feathers of dinosaurs.[4] However, Robert Dalgleish and his colleagues concluded that this specimen was not a louse,[5] thus leaving open the first discovery of a dinosaur-dwelling louse.)

The order Phthiraptera includes two major types of lice: Anoplura, or sucking lice, which feed only on mammals, and Mallophaga, or chewing lice, which feed on birds and mammals. Sucking lice have mouthparts that allow them to pierce their host's skin and suck its blood. If you have a child who has been sent home from school with a note alerting you to an outbreak of head lice, you are familiar with Anoplura.

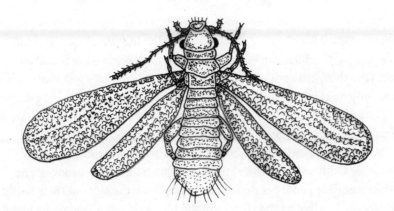

A

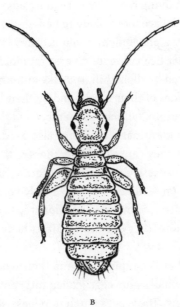

B

FIGURE 2.1   Booklice (Psocidae), likely ancestors of parasitic lice. (A) A fossil of a winged liposcelid from the mid-Cretaceous (about 100 million years ago). (B) An extant wingless booklouse (*Liposcelis corrodens*).

Mallophagans, which comprise about 85% of all lice, have mouthparts that are adapted for chewing. Zoologists subdivide mallophagan lice into three groups, of which two—Amblycera and Ischnocera—are found on birds. (The third group of lice, Rhynchophthirina, includes just three species, found only on elephants and warthogs. This group is likely a sister group of Anoplura, the sucking lice—meaning that they are derived from an immediate common ancestor and are therefore Anoplura's closest relatives.)

Amblyceran lice include about 1,350 species that are distinguished from other feather lice by their antennae, which lie within grooves, and by the sensory appendages that extend away from the sides of their heads. Amblyceran lice live on the surface of the bird's skin and feed on its blood by chewing at the base of newly emerging feathers, which causes itching and dermatitis for the unfortunate bird. Amblycerans are fast and agile, and they will try to desert a dying host. They are more adventurous than their ischnoceran relatives and are more likely to be found on a bird species that is not their typical host, a phenomenon that is called "straggling."

Ischnoceran feather lice are more diverse (at more than 3,000 species), and are easily recognized by their thin, rod-like antennae, which extend freely away from the sides of their heads, and by their lack of maxillary palps. They chew on feathers (which are metabolized thanks to bacteria that make an enzyme called keratinase), and they also eat debris at the surface of the skin. The damage they cause to the feathers of their host bird is highly visible, especially to prospective mates. Ischnocerans have bodies that are specialized for moving on feathers, and they rarely venture onto the bird's skin. When their host dies, ischnocerans seldom try to abandon ship—they are homebodies. Therefore, they depend on direct body contact with other birds for transmission.

Both types of feather lice typically spend their whole lives on the surface of their host. Females glue eggs (called nits) onto feathers to help the nits survive preening. These eggs hatch to release small, light-coloured nymphs. The nymphs feed and moult through three stages (called instars) to reach adulthood. The life cycle, from egg to egg-laying adult female, takes about fifty to sixty days, although this estimate is based on data from relatively few species.[6] In lab cultures, adults can live for about twenty to thirty

TABLE 2.1.   Mallophagan lice (Phthiraptera) found on birds

	No. of Species	Geographical distribution	Host birds
AMBLYCERA			
Boopiidae	1	Australia	Casuariiformes (e.g., emus, cassowaries)
Laemobothriidae	14	Worldwide	Charadriiformes (e.g., shorebirds), Ciconiiformes (e.g., storks, herons), Falconiformes (e.g., raptors), Galliformes (e.g., grouse, pheasants), Gruiformes (e.g., coots, cranes)
Menoponidae	650	Worldwide	Many orders
Ricinidae	65	Worldwide	Passeriformes (e.g., songbirds)
ISCHNOCERA			
Philopteridae	1,460	Worldwide	Many orders

Source: Marshall, A.G. (1981). *The ecology of ectoparasitic insects.* New York: Academic Press.

days, but here again, the data are sparse. There is some uncertainty about the lifespan of feather lice, because it is difficult to observe them over time in their natural habitat (on birds) and challenging to keep them alive for long off a host.

The host bird is a louse's entire world. Because lice spend their lives on the surface of one animal, and because new generations of lice are transferred to the young of the same host species, zoologists have reasoned that the evolutionary fates of both partners must be intimately linked. It makes sense for lice to have evolved specialized structural and physiological adaptations to help them survive and thrive in their habitat—that is, on their particular host. But one consequence of this specialization is that if a species of host goes extinct, its resident fauna of lice may also be doomed. Moreover, a successful adaptation by a species of lice may trigger a counter-adaptation by its host, initiating a cascade of changes to louse and bird alike.

When two different kinds of organisms affect each other's evolutionary histories—as happens, for example, with flowering plants and their animal pollinators—biologists call this co-evolution. Certain conditions

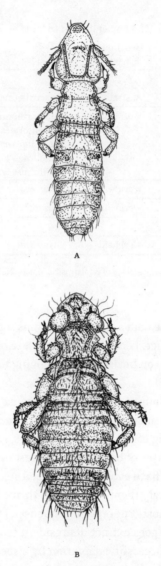

A

B

FIGURE 2.2   General features of amblyceran and ishnoceran lice. (A) The ischnoceran
louse *Aquanirmus bucomfishi* from a horned grebe (*Podiceps auritus*). Note the free antennae.
Ischnoceran lice are usually feather-eaters. (B) The amblyceran louse *Trinoton anserinum* from a
mallard duck (*Anas platyrhychos*). Its small antennae are hidden inside grooves on the underside
of the head. Amblycerans feed on blood and skin as well as feathers.

must be met for co-evolution to occur.[7] For instance, both partners must come from populations with some genetic variability; if all individual animals in a species were identical, there would be nothing for natural selection to select against. In order for hosts to evolve protections against parasites, it is important that they vary in some way. For example, individual hosts may vary in their immune defenses against parasites or in their ability to control parasites by preening. Parasites in turn must vary in characteristics such as their ability to escape and avoid the host's defenses or in their virulence to the host—that is, the likelihood of causing stress, damage, or illness to the bird. Usually, it is advantageous for lice to minimize damage, allowing their host to live long and prosper and to produce lots of chicks as future hosts.

An additional requirement for co-evolution is that both partners must encounter each other frequently or over a long period of time. Sporadic encounters aren't enough to trigger evolutionary changes. (In fact, if hosts become infected by new parasites, then the *lack* of a shared evolutionary history can lead to instability in the host–parasite relationship; the parasite may drive the host to extinction, or the host may not be a good environment for parasite reproduction.) In addition, both partners must exert some selection pressure *against* each other (or in mutualistic interactions, *for* each other) to trigger co-evolution.

The selection pressures that lice exert on their hosts include itching and dermatitis, physical damage to feathers, changes to the host's behaviour that drain its energy budget, and the transmission of infectious diseases or other animal parasites. Selection pressures that birds exert on lice include immunological responses by the host, moulting of feathers, preening, scratching, and a variety of other behaviours that aim at eliminating lice.

Because co-evolution is an ongoing, historical process, zoologists can explore it by looking for patterns—specifically, patterns of co-speciation and co-adaptation. Co-speciation occurs when the branching of parasites into different species mirrors that of their hosts as a result of the mutual adaptations between host and parasite. To find evidence of co-speciation, zoologists determine the most likely evolutionary history, or phylogeny, of a group of hosts, and compare it to the most likely phylogeny of their parasites. A perfect pattern of co-speciation would show a direct correspondence such that each species of host would have its own species of louse. The common occurrence of perfect or near-perfect co-speciation of host birds and their lice would have exciting implications because it would allow us to figure out the most likely evolutionary history (and the closest relatives) of a group of birds simply by looking at their lice.

Co-adaptation is the process by which interacting species undergo changes in lockstep with each other. If co-adaptation is common among birds and lice, then it should be possible to find some predictable patterns in the characteristics of lice by looking at characteristics of their host birds, and vice versa.

In fact, zoologists have found evidence of a number of patterns that arise from co-speciation and co-adaptation. Some of these patterns have been formalized as "rules," as stated in the following generalizations:[8]

> FAHRENHOLZ'S RULE: *There should be congruence between the evolutionary history (or phylogeny) of hosts and parasites.* Another way of expressing this is that the common ancestors of modern parasites were parasites of the common ancestors of modern hosts.
>
> EICHLER'S RULE: *A large taxonomic group of hosts (for example a family with many genera and species) should have more genera and species of parasites than a small taxonomic group.*
>
> SZIDAT'S RULE: *More recent or specialized host groups should have more recent or specialized parasites while more primitive or generalized hosts should have more primitive or generalized parasites.*
>
> MANTER'S RULE: *Parasites should speciate more slowly than their hosts.* When hosts diverge into separate species, there is a lag before adaptive changes are reflected in lice. Sometimes, the adaptations lice have undergone for living on an ancestral bird species will be adequate for life on the descendants. Thus, one order of bird could have many genera of lice. If evolution and speciation of lice were faster than bird evolution, a host would be expected to be infested with more than one species of a genus of lice.
>
> HARRISON'S RULE: *Large-bodied species of hosts should have large-bodied parasites.*

These "rules" describe some elegant patterns, but they are oversimplifications. In reality, they harbour many exceptions, because the co-evolutionary patterns that these rules describe are themselves intertwined with other patterns. But if one has the tools to interpret them, the exceptions to the rules can be just as illuminating as the rules themselves. For example, they may offer hints about the biogeographic patterns of hosts, such as their likely places of origin or changes to drainage basins that occurred in the host's history. Finding similar lice on host birds that are obviously not closely related tells us that the hosts must often interact in the same habitat—for example, cuckoos are often infested with lice from birds in whose nests they lay their eggs. Species of lice that resemble each other, but turn out not to be recently related in an evolutionary sense, may also tell us a lot about the kinds of selection pressures that affect birds and their parasites. Furthermore, looking closely at the exceptions to the broader patterns may provide insights into the length of time that a parasite has been associated with a particular host, and therefore, whether there has been enough time for co-evolution to have occurred. Such exceptions may contain important clues, such as the degree of host specificity of a louse (the more exclusive the louse is in its habitat, the longer it is likely to have inhabited that host), or the severity of the effects of the parasite on its host (hosts that have a long history with a particular parasite tend to evolve biological defenses against them). The number of interesting questions and new knowledge generated by these patterns and their deviations is staggering.

INTERTWINED ANCESTRIES

On a large scale, there is good evidence of co-evolution and, more specifically, co-speciation. For example, orders of birds often have families and genera of lice that do not occur on other birds. Such correlations have led zoologists to reason that if a population of hosts somehow became genetically isolated from other members of their species and over time formed a new species, then their lice would also either change (to form a new species) or go extinct. Cases of host switching, where one species of feather louse becomes established on an unrelated species of host that occurs in the same habitat, should therefore be rare.

Thus, it was predicted that if a species of host went extinct, like the passenger pigeon (*Ectopistes migratorius*), its resident lice would also be wiped out. Hence, when a species of feather louse, *Columbicola extinctus*, was found on a stuffed passenger pigeon in 1937, 23 years after its host went extinct, it was assumed to be among the last of its kind. Surprisingly, the resurrected louse was later found alive and infecting band-tailed pigeons (*Patagioenas fasciata*) in 1999.[9] Band-tailed pigeons are close relatives of passenger pigeons, so a plausible story is that lice infested a common ancestor of both species of pigeons, and managed to survive on band-tailed pigeons after the extinction of passenger pigeons, thereby escaping the evolutionary fate of the doomed pigeon hosts.

Assumptions about the tightly correlated histories of lice and hosts led some early zoologists to be so confident of co-speciation that, whenever they found lice on a bird that had not been examined previously, they decided the lice must represent a new species, simply because it was on a new host. This has kept modern zoologists who work on the taxonomy of lice busy—most often by debunking many early-described "species" and more accurately lumping several former species into one.

With more-detailed study, the relationship between species of lice and species of hosts has turned out to be less clear than originally thought, but more interesting. For example, Dale Clayton and Kevin Johnson looked at two kinds of ischnoceran feather lice found on doves (Columbiformes).[10] One type, which Clayton and Johnson call body lice, are found on the host's abdominal feathers and are represented by the genus *Physconilloides*. These lice have stubby bodies and try to escape preening by hunkering down into the downy parts of feathers. They are very host-specific (usually one species occurs on one host species) and different genetic races occur on geographically separated hosts. Analysis of mitochondrial and nuclear DNA of both body lice and doves showed that their phylogenies matched in 8 out of 12 evolutionary events. This reflects a high degree of co-speciation.

But a different story has unfolded for feather lice of another type, also found on doves. These lice (represented by the genus *Columbicola*) are ecologically different. They occur on the flight feathers of doves' wings and tails—Clayton and Johnson call them wing lice. They have long bodies and try to escape preening by wedging themselves into the spaces between feather barbs. They are less host-specific and genetically less diverse. Wing lice have demonstrated much less evidence of co-speciation—alignment occurred in only 3 of 12 events.

Why is the evolutionary trajectory of wing lice so much less intertwined with their hosts than that of their distant cousins, the body lice? Clayton and Johnson attributed the difference in co-speciation between body and wing lice to the difference in their habitats, ecology, and behaviours—especially their ability to travel.

Although both types of lice are transmitted from dove to dove by direct body contact, feather lice can also hitch rides on parasitic flies (of the order Diptera) called louse flies (Hippoboscidae). Louse flies do not tend to be very host-specific and can carry lice to new bird hosts. Zoologists call this hitch-hiking phoresy. Preliminary evidence suggests that wing lice are much more likely to be hitchhikers than body lice. Clayton and Johnson believe that their looser co-evolutionary relationship with doves was directly affected by their nature as adventurous fortune-hunters. However, this characteristic can in turn be explained by several other factors.

Phoresy is a dangerous strategy. Lice may be groomed off by the louse fly, may fall off during travel, or may end up on a very unsuitable bird. In an experimental study, Andrew Bartlow and his colleagues found a trade-off between louse mobility and phoresy—the least mobile lice were the most likely to take the risk of using louse flies, presumably because they had so few options for getting from one host to another.[11] However, the researchers also suggested that competition between different species of lice on one bird may have caused some lice to resort to phoresy in hopes of finding less populated territories to colonize. These studies all suggest that exceptions to the general pattern of co-speciation may turn out to be far from random. In fact, they may reveal other intricate patterns between a louse and its host environment.

Another interesting louse–bird co-evolutionary study was reported by Adrian Paterson and Russell Gray, who analyzed the relationships between seabirds (albatrosses, petrels, penguins, and cormorants) and their feather

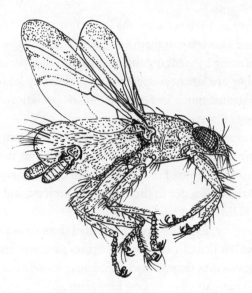

FIGURE 2.3 Two feather lice hitching a ride on a hippoboscid louse fly. This process of transporting lice from one bird to another is called phoresy.

lice.[12] They used several methods to reconstruct detailed evolutionary diagrams for fifteen species of seabirds and their lice, and then compared how well the diagrams for the lice matched those of their hosts. Using a detailed statistical analysis, they concluded that the relationship between the lice and the seabirds pointed to "a consistent history of co-evolution." Based on this strong evidence of co-evolution, Paterson and Gray were able to make new predictions about the occurrence of certain lice on specific hosts—for example, they predicted that zoologists should be able to find *Epsibates* lice on mollymawks (Buller's albatross, *Diomedea bulleri*) and *Naubates* lice on sooty shearwaters (*Puffinus griseus*), because these kinds of lice occurred on the birds' closest relatives. These predictions were later borne out.

Still, the correspondence between lice and their hosts was far from perfect. For example, two genera of lice (*Austrogoniodes* and *Episbates*) occurred on four species of penguins but—surprisingly—also infested the royal albatross (*Diomedea epomorphora*), a species that was nowhere near the penguins in the evolutionary tree. Such surprises prompt zoologists to speculate about their causes. In this case, host switching seems a likely explanation—and, because seabirds tend to be colonial birds, and therefore have little opportunity for contact with other types of birds, the switching likely happened via phoresy, as there would otherwise be little opportunity for lice to switch from one bird species to another.

In the case just described, the surprise involves unexpectedly finding a louse on a host where we would not expect it to occur based on the evolutionary histories of the birds and lice. But in other cases noted by Paterson and Gray, certain lice failed to show up on expected hosts. How can zoologists explain this absence, given the otherwise strong evidence for co-evolution? Paterson and Gray proposed that these gaps in the tree were due to "sorting events"—that is, various events that resulted in the disappearance of lice from their hosts. For example, suppose that one species of bird, carrying a specific louse species, diverges into two separate species. The original louse, perhaps unable to adapt to the traits of the new version of its host, goes extinct, leaving an evolutionary tree that is no longer perfectly matched with that of its host. Or, suppose that several individuals of a bird species relocate and establish a new and successful bird population, over time evolving into a slightly different species from their colony of origin. And suppose, moreover, that the intrepid founding birds happened to have been especially successful at avoiding being infested by one of the lice that commonly afflicted their original colony. These lice will have "missed the boat" to the new colony, and therefore will be absent from that entire branch of the host's evolutionary tree.

Fortunately, there are methods for evaluating the explanations that arise out of zoologists' attempts to account for surprising mismatches between hosts and louse evolutionary histories. For example, a gene in the DNA of mitochondria of animals incurs mutations that allow it to be used as a molecular clock. By comparing differences in the genes of two different species of animals (lice or birds), we can figure out how long ago the two species diverged from a common ancestor. This provides valuable information for reconstructing the events that caused mismatches in the evolutionary trees of lice and their hosts.

In other cases, the unexpected occurrence of certain lice on a bird may hint at an ecological explanation. For instance, *Strigophilus* feather lice occur only on owls, and are often quite host-specific. Dale Clayton examined a group of lice in this genus called the cursitans group, and found that 17 of 24 lice in this group each inhabit a single species of owl; two occur on two or more owls that belong to the same genus; however, the other five can be found on a variety of owls that may be distantly related to each other.[13] For example, the far-ranging *Strigophilus syrnii* infests not only great horned owls (*Bubo virginianus*) in the New World, but also great gray owls (*Strix nebulosa*), spotted owls (*Strix occidentalis*), barred owls (*Strix varia*), and even South American rufous-barred owls (*Strix albitarsus*). As it happens, great horned owls are very catholic with regards to habitat and overlap ranges with the other species of owls. All of these species nest in holes or in the abandoned nests of other birds, so opportunities for sharing lice are great. In fact, among the cursitans group, any lice that were found to reside on more than one type of owl were without exception found on owls that overlapped in range and shared similar habitats. Clayton concluded that finding a shared species of louse on more than one species of bird does not necessarily indicate that hosts are evolutionarily close species, but can tell us something about their ecology—in this case, their orientation toward setting up housekeeping in "time-share" habitats.

Brood parasites such as cowbirds and cuckoos, which are famous for laying their eggs in the nests of other birds, may expose their offspring to the parasites living on the foster parents who incubate, feed, and care for the parasitic nestlings (see more about brood parasites in Chapter 9). A study of brown-headed cowbirds (*Molothrus ater*) found that adult birds were infested by 6 species of lice from 5 different genera, and fledglings were infested with 11 species from 6 genera, mostly amblycerans.[14] These diverse "cowbird" lice could be attributed to a variety of passerines, especially red-winged blackbirds (*Agelaius phoeniceus*) and wood thrushes (*Hylocichla mustelina*). In fact, examination of lice on cowbirds allowed researchers to determine the identity of the foster parents that had raised them—an interesting example of nurture versus nature.

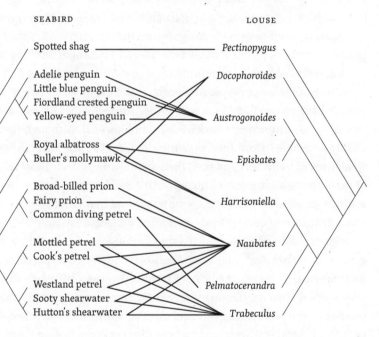

SEABIRD	LOUSE

Spotted shag ———————— *Pectinopygus*

Adelie penguin
Little blue penguin
Fiordland crested penguin
Yellow-eyed penguin

Docophoroides

Austrogonoides

Royal albatross
Buller's mollymawk

Episbates

Broad-billed prion
Fairy prion
Common diving petrel

Harrisoniella

Mottled petrel
Cook's petrel

Naubates

Westland petrel
Sooty shearwater
Hutton's shearwater

Pelmatocerandra

Trabeculus

FIGURE 2.4   The pattern of co-evolution between seabirds and their lice. The evolutionary history (phylogeny) of the birds is based on behavioural, protein, DNA, and RNA data while the phylogeny of lice is based on morphology (body size, shape, and features) and life history data. The illustrated bird is Buller's Mollymawk (also called Buller's albatross), *Diomedea bulleri*, and the louse is from the genus *Docophoroides*.

Similarly, studies of cuckoos in Africa and Japan revealed that adults have three or four species of lice and mites that are specific to cuckoos but, as with the cowbird study, passerine lice from foster parents were also found to infest cuckoo fledglings. Lice that are specific to cuckoos are acquired later, after fledging, by body contact between adults.

The moral of the story is that lice can provide insights into their hosts' likely evolutionary history, but that the rules of co-evolution should be applied with caution, and ecological factors need to be considered. The evolutionary story of flamingos drives this lesson home in an especially forceful way. Long ago, Miriam Rothschild and Theresa Clay suggested that lice could be used to provide insights into the ancestry and closest relatives of flamingos (order Phoenicopteridae).[15] Flamingos are long-legged, wading filter feeders. Determining their place in evolutionary history has been "one of the most perplexing problems in avian systematics."[16] Several studies have placed flamingos with either Anseriformes (ducks, geese, and swans) or Ciconiiformes (storks, ibises, and herons). Although flamingos closely resemble Ciconiiformes in their external appearance, the specific structures of their bills and feet, their vocalizations, and the development of their young all suggest a kinship with Anseriformes. A third hypothesis was put forward by D.S. Peters, who suggested that flamingos are related to Charadriiformes (wading birds, including gulls and terns).[17]

Rothschild and Clay reported that flamingos are infested by three genera of feather lice (*Anatoecus*, *Anaticola*, and *Trinoton*) found only on Anseriformes, whereas the lice of Ciconiiformes are different. The conclusion seemed obvious: flamingos and geese came from a common ancestor. But recent molecular evidence, coupled with a more-detailed analysis of their physical features, led to a very different conclusion: flamingos are actually related to a group of birds that no one had ever considered—grebes (Podicipediformes).[18]

Rather than flamingos evolving from ancestral birds that frequented the shore and then evolved to live on water, it appears flamingo–grebe ancestors evolved from the water to the shore, and may have acquired lice from ducks, geese, and swans along the way (interestingly, this conclusion is supported by another type of parasite, a family of tapeworms, which will be discussed in Chapter 7). This possibility led to a reinterpretation of the similarity in lice faunas between flamingos and ducks by Kevin Johnson and his colleagues.[19] These researchers concluded that the lice of grebes and flamingos are actually closely related in evolutionary history, whereas the similarity between lice found on ducks and flamingos is more superficial and due to host switching. Perhaps even more surprising, they believe that ducks acquired their lice from flamingos, rather than the other way around.

The story of flamingos and their lice is a complicated one. In other studies (for example, with ostriches, rheas, and emus), lice have been excellent and more direct witnesses, providing reliable evidence about the evolution of these hosts.

The best conclusion is that the "rules" describing the relationships of feather lice and bird hosts are excellent working hypotheses—useful tools for biologists to formulate testable questions. Detailed studies that consider these general patterns along with more specific factors, such as the ecological traits of louse and bird, can provide wonderful insights about this section of the flying zoo.

PREENING AND ITS DANGERS

The strength of the evidence for co-evolution between birds and lice suggests that selection pressures exerted on both partners must be formidable. For years, people have observed that birds with deformed beaks, or birds that were prevented by scientists from being able to preen normally, suffered from massive outbreaks of lice, at times housing more than ten times the normal number of lice. In one such study, scientists inserted metal bits into the beaks of rock doves (*Columba livia*) to disrupt their preening hygiene. As a result, their louse populations exploded, and the birds lost up to 30% of their feathers, had higher metabolic rates due to lost insulation from the damaged feathers, and continuously lost body mass.[20] Uncontrolled, lice can exert a great cost on their hosts.

Birds normally spend about 10% of their time grooming (mostly preening) and scratching, but birds that host more species of lice spend more time preening than those that host fewer.[21] Dealing with parasites incurs great expenditures of precious energy, energy that could be used instead for growth and reproduction. The sizes and shapes of birds' bills have long been recognized as important for foraging and feeding, but their use as tools for removing ectoparasites—that is, parasites that live on the outside of the host's body—was not fully appreciated until the 1990s.

Dale Clayton and Peter Cotgreave found that birds such as toucans, which are fruit-eaters with long, unwieldy bills, are less adept at removing ectoparasites than their shorter-billed relatives, the aracaris.[22] In fact, deprived of an effective tool for preening, long-billed birds resort to spending much more time scratching than preening. As demonstrated by one study, the bill's competing functions often clash; the extent of the overhang of the upper bill of birds is directly correlated with their ability to remove feather lice, but preening ability comes at the price of foraging efficiency.[23]

To a louse, the host's beak is the most fearsome predator in its environment. As a result, lice have developed the ability to hide in sites—or microhabitats—that are difficult for birds to preen. To compare the vulnerability of different microhabitats to preening, Lajos Rozsa painstakingly glued dead feather lice (*Columbicola columbae*) on either the underwing covert feathers or the tail feathers of pigeons (*Columba livia*). After 24 hours, half of the lice on the wings had been removed, whereas almost none were removed from the tail.[24] Lice instinctively seek out safer locations; for example, they often migrate from more dangerous sites to the head (which is impossible for birds to preen) to lay their eggs.

Some lice have become specialized for life in other microhabitats where eviction is difficult. For example, the large (1.5 cm) louse, *Piagetiella perialis*, lives inside the mouth and gular pouches of the white pelican (*Pelecanus erythrorhynchos*), where it feeds on blood. Bill Samuel and his colleagues reported infestations of more than 300 lice on some pelicans, causing ulcers in their mouths. They saw a white pelican hatchling become infested with its first louse from a nestmate, even before it was free of its own eggshell.[25] Another amblyceran louse, *Actornithophilus*, lives in an equally inaccessible site: inside the quills of wing feathers of curlews, *Numenius americanus*.[26]

Even when populations of feather lice on birds are small, as they often are, they can still play an important role in exerting selection pressures, because lice transmit other parasites whose effects may be severe. This can be seen in red-necked grebes (*Podiceps grisegena*), diving waterbirds that blast through the water using their powerful legs and webbed toes. These birds often have legs with large, swollen "knees"—actually, the intertarsalis joint, or ankle. Usually, only one leg is affected but it is not uncommon to see grebes with "football player knees" on both legs. Dissection shows that the swollen joints are filled with yellow fluid containing nematodes, or roundworms (*Pelecitis fulicaeatra*). The worms cause inflammation and displacement of the Achilles tendon, leading to tendonitis, which is the cause of the swelling. The grebes become infected with these worms thanks to amblyceran lice (*Pseudomenopon dolium*) that inject the worms' larvae into their skin or bloodstream.[27] Although no studies have directly compared the behaviours of infected and uninfected grebes, we can imagine that the nematodes cause pain, thereby reducing mobility and limiting an infected grebe's ability to forage efficiently.

The benefits and dangers of preening have therefore led to a high degree of co-adaptation between lice and their hosts. This strong co-adaptation is plainly reflected in the observation that the colour and size of lice are related to the colour and size of their hosts. Black swans are infested with black lice, and white swans with white. This camouflage helps body and wing lice escape preening. However, for bird head lice, which can escape preening, colour coordination with host plumage is not as important, unless the host birds are in the habit of allogrooming, or grooming each other.[28]

The relationship between the body size of lice and the body size of their hosts has been recognized as Harrison's Rule. The idea is that lice on smaller birds have to be adapted to live in smaller feathers, because large lice on small birds would be detected by the host and vulnerable to grooming. This rule was examined in a study by Kevin Johnson and his colleagues.[29] They found that the sizes of 78 species of amblyceran and ischnoceran lice, taken from a variety of different birds, were significantly related to the body masses of their hosts. However, as we saw in the previous section, the "rules" of co-evolution are subject to exceptions that typically point to more

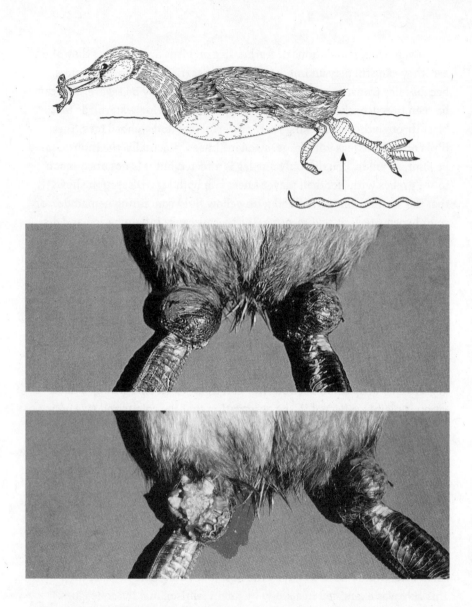

FIGURE 2.5 "Football player knees"—actually the swollen tibio-metatarsal joint—in a red-necked grebe (*Podiceps grisegena*) caused by a nematode roundworm (*Pelecitus fulicaeatrae*), which is transmitted by lice. Red-necked grebes feed by using their legs and lobed toes to propel themselves underwater, but the pathogenic effect of the worms is unknown. Lice acquire the infection by feeding on bird blood that contains microfilaria (larvae).

nuanced explanations. Indeed, a more in-depth analysis of lice specific to doves showed that Harrison's Rule was generally supported for 19 species of wing lice, but not for 24 species of body lice. Because it is easy for birds to preen their wings, there is extreme pressure on wing lice to match the size of their host. Body lice, which can live in hard-to-preen places on the head and around the eyes, may not be subject to the same pressure. However, the relative safety of their habitat means that body lice may be forced to contend with competing species of lice that have set up housekeeping on the same bird—it is not unusual for body lice on a dove to have to compete with one or more other species on their host. As a result, body lice can be driven out of microhabitats where they are relatively safe from preening, and into microhabitats where they are now under pressure to match the size of their host (especially when body size scales to the space between feather barbs). This dynamic can account for some interesting patterns in the size of body lice. Large lice make fierce competitors, and are better able to drive other lice out of the best microhabitats. However, they also have fewer places to hide on a small bird, which limits where on the bird they can safely live. Hence, it is less advantageous to be a large body louse on a small bird. The factors that determine the likely size of the body lice, then, are quite complex, involving the interaction of competition and preening. In contrast, the factors that determine the size of wing lice are much simpler.

Preening is a major selective force acting on lice, but it is not the only logical explanation for Harrison's Rule. Dale Clayton and his colleagues carried out experiments to see how much of the size effect was due to differences in the abilities of lice to escape preening or in their abilities to remain attached to their hosts or to feed.[30] They focused their study on dove lice (*Columbicola*) because there is evidence for co-speciation between *Columbicola* lice and doves and, more specifically, there is a strong correlation between doves' body size and the size of their lice.

To test the ability of dove lice to remain attached to different host species during flight, the lice were placed on feather grafts and subjected to simulated flight using a high-speed fan, or were transferred to a different species of host that flew on a tether. To test their feeding habits, captive lice were fed feathers from their normal host as well as feathers from other doves. Finally, Clayton and colleagues put lice on regular and irregular hosts, each with normal or blocked preening ability. The results conclusively demonstrated that lice could remain attached to and feed on a variety of doves (even if they differed in size by an order of magnitude), but they could not escape preening on hosts that differed in size from their normal host. Preening was the major selection pressure that led to size correlations between lice and birds. This is strong evidence that preening has played a major role in co-speciation.

Preening has also been an important factor affecting the microhabitat distributions, body shapes, and escape behaviours of lice. It has been known for a long time that feather lice sort themselves on the body of their host in a predictable fashion. For example, domestic pigeons are infested by four species of feather lice, two ischnocerans and two amblycerans. The ischnocerans are *Campanulotes compar* (found mostly on the neck, back, sides, and breast) and *Columbicola columbae* (found mostly on the wings). The amblycerans are *Hohortstiella lata* (found scattered over the body) and *Colpocephalum turbinatum* (found mostly on the wings and tail).

Researchers have found that all of these lice have similar tastes, eating the downy parts of feathers (although *C. turbinatum* is also cannibalistic, and *H. lata*, as an amblyceran, also feeds on blood). But they have different strategies for escaping preening.[31] *C. compar* is a small, slow louse that lives near the bottom surface of the feather rachis. When the feather upon which it is feeding is disturbed, this louse drops onto another feather. *C. columbae* is large and slender—well adapted to rest in the furrows of the undercoverts of the flight feathers. When disturbed, it grabs hold of the feather barbules with its mouthparts and tries to weather the storm. *H. lata* is a large, stout louse that rapidly runs over the skin surface, and *C. turbinatum* is a small, fast louse that travels along a feather's quill and in furrows between the barbs of primary and secondary feathers, trying to outrun its predator, the host's beak. All of these lice also lay their eggs in predictable but different locations on a pigeon so that the nits have the best chances of avoiding preening. Hence, the researchers have concluded that preening is the main selective pressure acting on these lice. Their characteristic behaviours, their living quarters, their body shapes, sizes, and proportions—all are the result of their host's relentless preening, in the ever-escalating co-evolutionary war between lice and birds.

As lice have adapted their bodies and behaviours to mitigate the effectiveness of preening, birds have countered these tactics by employing yet other weapons. Some birds preen each other (zoologists call this allopreening), although this technique is used more often for removing ticks than for lice. Allopreening occurs mostly between mutually consenting adult birds, although parental preening of young has also been noted. Its importance for removing ectoparasites is debated, however, and is likely most significant for removing lice that have escaped to the head and neck of their host, where self-preening is difficult.[32]

Occasionally, birds use other organisms as anti-lice weapons. Naturalists have seen birds from more than two hundred species (mostly passerines) lying on ant nests, or in some cases, picking up ants and rubbing them onto their feathers—a phenomenon called "anting." Many ants secrete formic acid and chemicals called terpenoids, which may not only repel and kill feather lice, but also protect birds from fungal and bacterial skin infections.

Common grackles (*Quiscalus quisqula*) have been observed "anting" in marigold leaves. As many gardeners know, marigolds contain pyrethrin (an insecticide) and other compounds, such as sitosterol, that can affect the reproduction of arthropods. Marigolds are often planted as companions to more delicate flowers and vegetables to protect them from insect pests. Perhaps grackles are effectively spraying themselves with insecticide.

Besides employing natural plant products, urban birds sometimes include discarded cigarette butts in their nests. This chemical warfare likely helps to control ectoparasites—especially those that might infest helpless nestlings.[33]

A more direct method is used by three species of brightly coloured tropical rainforest birds in New Guinea, belonging to the genus *Pitohui*. These birds use their own bodies to poison lice. Pitohuis carry a neurotoxin (homobatracotoxin—the same kind of toxin found in poison dart frogs—in their skin and feathers). Experiments have shown that the toxic feathers of *Pitohui dichrous* are avoided by lice, and for good reason—it kills them.[34]

Other tropical birds, such as South American hoatzins (*Opisthocomus hoazin*), eat leaves of rainforest trees that contain distasteful alkaloids. People in the Amazon region of Ecuador call hoatzins "stinky turkeys" because of the birds' bad taste, likely a result of their diet. It would be interesting to study the diets of these birds in more detail; besides deterring predators, hoatzins may gain some protection from parasites as a result of their distasteful feathers. The lice that do reside on them may be uniquely adapted to eat them, a notion that is bolstered by strong evidence for co-evolution between these hosts and their lice. Hoatzins are famous in the ornithology world because young birds have claws on their wings that are used to scramble back up into the branches of trees when the birds fall out of nests. This feature suggests that hoatzins are very primitive, an idea that has been supported by molecular evidence.[35] Six species of lice have been reported to live on hoatzins.[36] Four of these lice are known only from hoatzins (*Osculotes curta, O. macropoda, Hoazineus armiferus,* and *Pessoaiella* (= *Wilsonia*) *absita*), evidence of the long co-evolutionary history between lice and their hosts. (The other two usually occur on cuckoos and falcons—these are probably accidental stragglers, picked up because of cuckoos' nest parasitism and shared habitat.) It would be interesting to know what allows these lice to tolerate what is usually a repellent diet.

The chemical weapons just described are specialized to individual species, not specifically targeted at lice, and probably not very common. But all birds have immune systems. Immunity provides birds with protection from specific enemies, but triggering it requires direct contact between parasites and living host tissues.

Anders Moller and Lajos Rozsa reasoned that amblyceran lice, which live and defecate on bird skin surfaces and feed on blood, should induce immune reactions in birds, whereas ischnocerans, which graze peacefully on non-living, keratinized feathers, should not. In a study of many species of birds, these researchers found that birds infested with a large variety of amblyceran lice had stronger immune responses than those who hosted fewer types, whereas there was no relationship between immunity and the number of ischnocerans.[37] In fact, Moller and Rozsa argued that the specificity of the immune response has promoted a diversity of amblyceran species by accelerating the evolution of new amblyceran species that are able to counteract their host's immune defenses. Thus, while behavioural defenses like preening and anting have encouraged co-speciation and co-adaptation among all lice, immune reactions could have contributed strongly to the evolution of amblyceran lice in particular.

IN THE GALAPAGOS LAB

Weaving all of these interacting evolutionary and ecological patterns together into one coherent story is difficult. Perhaps we can get a more concrete sense of how they play out by considering a specific example. And where better to find that example than in the living lab of evolution—the Galapagos Islands?

The Galapagos Islands (made famous by Charles Darwin for his close study of birds there) are recent volcanic islands, situated on the equator. The top predator (and scavenger) in the islands is the endemic Galapagos hawk (*Buteo galapagoensis*). Galapagos hawks are most closely related to— and probably derived from—Swainson's hawks (*Buteo swainsoni*). Swainson's hawks are common in the prairies of North America, but they migrate, mostly over land, all the way to southern South America. It is thought that a storm once blew some hawks off course, and they founded a population in the Galapagos, later speciating to become Galapagos hawks.

Galapagos hawks probably survived in the islands because of their catholic diet. They have been seen feeding on young marine and land iguanas, lava lizards, young tortoises, all kinds of birds (especially mockingbirds, doves, and Darwin's finches), a variety of invertebrates, and the ever-present carrion (Darwin called them "carrion buzzards"). Today, the hawks nest on all major islands in the archipelago except Genovesa, which they probably failed to reach, and San Cristobal and Floreana, where they were eradicated by people.

Large female Galapagos hawks mate repeatedly (with each female taking part in five to ten copulations per day) with two or three smaller territorial males—although some females are monogamous, while others have "harems" of up to eight males. The males engage in cooperative polyandry, which means that they participate in incubating eggs, feeding, and raising young, and together, they vigorously defend a territory. Roosting together, males show no signs of jealousy and male–male copulations have been observed.

Usually one to three chicks hatch, and young hawks are expelled from the nest at about four months of age. Young adult hawks spend two or more years in non-territorial areas before they get the opportunity to breed.

Galapagos hawks are infested by two species of lice, *Colpocephalum turbinatum* (an amblyceran) and *Degeeriella regalis* (an ischnoceran). As usual, the amblyceran lice feed on epidermal tissues and blood, and are less restricted to particular microhabitats on hawks. They are vagabonds, and transmission between birds other than offspring is common. True to their wandering ways, *C. turbinatum* has been reported from 53 species of birds (although the Galapagos hawk is the only known host in the Galapagos Islands). The ischnoceran *D. regalis* is a feather-eater that is much more restricted to particular microhabitats on the host. In the New World, it has only been reported from two hosts, Galapagos hawks and, not surprisingly, Swainson's hawks.

Almost all (90% to 100%) of the Galapagos hawks on the islands of Marchena and Santiago were found to be infested with *C. turbinatum* while somewhat fewer hawks (70% to 100%) were infested with *D. regalis*.[38] On average, about four times as many amblycerans as ischnocerans occurred on male hawks, averaging about 10 to 47 amblycerans per bird in contrast with 2 to 10 ischnocerans per bird).

Among territorial male hawks, members of larger groups (four to six birds) were infested by significantly more amblycerans than males from smaller social groups (one to three birds). But while the degree of sociality affected *Colpocephalum*, this trend did not occur for *Degeeriella*. This contrast is probably due to differences in the transmission of the two kinds of lice. Amblycerans are more likely to be transmitted horizontally between adult hosts, while ischnocerans are transmitted vertically from parents to chicks. Group size has been shown to be associated with greater, and more diverse, louse loads in other birds[39] and no wonder—lice can be transferred horizontally during copulation in as little as two seconds![40]

Not only did the numbers of *C. turbinatum* lice vary with group size, but their populations were distributed differently than those of *D. regalis*. The distribution of the latter was clumped, with a few birds having many lice and most hawks having few. For the amblyceran lice, however, there was not the same degree of clumping, again probably reflecting the high degree of transmission of these lice between male hawks. If lice can have an important effect on the fitness of hawks—for example, by spreading other parasites—their transmission patterns and population distributions could put downward pressure on the size of the birds' social groups, and may in fact be responsible for the average group size of two or three male hawks.

The health of Galapagos hawks is closely connected with the number of lice on their bodies, but it is not always easy to identify which of these is the cause and which is the effect. Noah Whiteman and Patricia Parker found that non-territorial males were in significantly worse body condition than territory holders, and also had more lice of both types—270 versus 26 for *C. turbinatum* and 22 versus 7 for *D. regalis*.[41] (When the number of amblycerans on a bird was greater than 100, there tended to be *fewer* ischnocerans, probably because *Colopcephalum* were eating *Degeeriella* or displacing them from their normal microhabitat to areas where they could be removed by preening.) In general, louse load and body condition of Galapagos hawks were good predictors of territory ownership, and thus bird breeding opportunities.

But this study leaves us with unanswered questions: Do the deleterious effects of lice on birds lead to poor body condition, and therefore, an inability to hold territories and fewer reproductive opportunities? Or does lack of territory lead to stress, poor nutrition, less time to spend on preening, poor immunity, and thus, more lice? Are lice a driver in this system or simply an indicator of other problems?

There is evidence that resistance to parasites, including lice (and especially amblycerans), is hereditary. It also seems likely that mate choice in birds is based, at least in part, on information about parasite resistance, as shown by intensity of colour, symmetry of feathers, condition of plumage, displays, and ability to hold a territory. Therefore, it is likely that lice, even in small numbers, can have pronounced effects on the health and condition of birds and can affect their Darwinian fitness. The co-evolutionary dance (including co-speciation and co-adaptation) goes on in a never-ending arms race that leads to an uneasy truce between birds and their feather lice.

In 1842, a zoologist named Henry Denny decided to make lice the subject of his life's work. He wrote: "In the progress of this work, however, the author has had to contend with repeated rebukes from his friends for entering upon the illustrations of insects whose very name was sufficient to create feelings of disgust."[42] Although most people look upon lice with disgust, the remarkable relationships that they have with birds—one of the best examples of evolution by natural selection in the natural world—leads me to conclude that lice are indeed beautiful!

3

F L E A S

The Circus in the Zoo

ADAPTING TO LIFE ON A BIRD

Most people regard lice with disgust but are fascinated by fleas and may even consider them to be cute—they are the panda bears of the flying zoo. Fleas have always captivated people. Aristophanes, in *The Clouds*, describes how Socrates and Chaerephon tried to measure the length of a flea's jump in proportion to its body size: They took a wax impression of a flea's foot and then scaled this to calculate the length of its body. Then they measured the distance of a flea's jump from Socrates's hand to Chaerephon's and resolved the problem by simple multiplication. They estimated that if a man could jump as athletically as a flea, he could leap one hundred metres high![1]

Due to their remarkable athletic ability, fleas were kept in "circuses" for the amusement of people right up to the 1940s. Circus fleas were fitted with saddles and harnesses, and were made to pull carriages, cannons, and other props around a stage. Fleas actually aren't stronger than other insects, but they are easy to collect—and since they can be fed on the arm of their owner, they survive well in captivity.

Fleas (order Siphonaptera; *siphon*—"a tube"; *aptera*—"without wings") are obligate blood feeders; as adults, they can live for long periods of time away from a host, but both males and females must have a blood meal, usually from a bird or mammal. Fleas likely evolved from insects similar to scorpionflies (order Mecoptera), particularly from a family known as snow scorpionflies (family Boreidae). Snow scorpionflies usually live in moss and are often found on snow in winter. Males have reduced, bristle-like wings and females have small, flap-like wings. Although scorpionflies have chewing mouthparts, these are elongated to form a tube-like snout, perhaps a precursor for blood feeding. Study of the DNA sequences of four genes, as well as evidence from their body shapes, suggests that fleas and snow scorpionflies are closely related sister groups.[2]

Fleas probably evolved as parasites of burrow-dwelling mammals, especially rodents (almost 75% of the 2,500 known species of fleas infest rodents); however, they later infested arboreal, nest-dwelling mammals like squirrels (accounting for about 20% of current flea species), and, most recently, birds (about 5% of current species). One species (*Ceratophyllus lunatus*) may have given up on birds and now infests martens and fishers, which are small carnivorous mammals.[3]

Today, fleas show many adaptations for life as ectoparasites—that is, parasites that live on the outside of their hosts' bodies. Their bodies are streamlined and flattened laterally, which allows them to move through feathers and hair with ease (the bodies of lice are also flattened, but dorsoventrally—from back to belly). The antennae of fleas lie in grooves, as do those of lice, so they don't catch on feather barbs or hair. And, like lice, they have no wings, although larvae do develop rudimentary wing buds, which suggests that they evolved from winged ancestors. The body usually features two rows of strong, backward-facing spines (called the pronotal and genal combs), made of a tough, horny protein called chitin. These spines protect vulnerable areas of the body where the cuticle is thin, reduce the loss of fluids to delay desiccation, and they may also help the flea attach to its host.

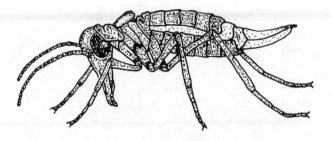

FIGURE 3.1 A female snow scorpionfly, *Boreus brumalis*. Note the small, flap-like wings. In spite of their fierce-sounding name, scorpionflies are harmless. Insects resembling these may have been the ancestors of fleas.

Interestingly, bird fleas belonging to four different families all demonstrate similar changes to the pronotal comb over the course of their evolution.[4] The spines of the comb have increased from 24 (on the mammal fleas) to 36. The spines are narrower and the fleas have a broader girth so that the comb will fit comfortably. In addition, on all four types of bird fleas, the spines are close-fitting, parallel-sided, and have a horizontal position. Many other hairlike spines called setae are distributed on the body—perhaps for protection and also to provide touch-based sensory information. On bird fleas, these bristles are longer and more delicate than the same bristles on related mammal fleas. These modifications, which are strikingly similar across species of bird fleas, probably represent convergent evolution—that is, they likely represent special adaptations that have evolved quickly but independently in each of the four flea families to allow them to inhabit a feathery—instead of a hairy—environment.

Given that fleas live in dark environments such as burrows, nests, and fur pelts, they have simple eyes rather than compound ones; however, the eyes of fleas that attack diurnal hosts are better developed than fleas on animals that are nocturnal or use burrows. There are many other sensory receptors that detect chemicals (especially carbon dioxide), vibrations, air currents, and even ultrasonic sounds that fleas probably use to communicate with each other—imagine singing fleas! Both sexes have a saddle-shaped

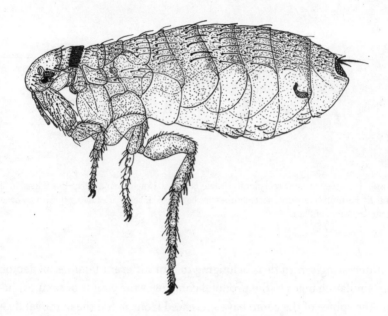

FIGURE 3.2   A female swallow flea, *Ceratophyllus idius*. This flea is found in Alaska, across northern Canada to Labrador and Newfoundland, and in the eastern part of the United States. It occurs in swallows' nests, especially those of tree swallows (*Tachycineta bicolor*). Bird fleas show many structural adaptations for ectoparasitism: antennae are small and housed in grooves; the pronotal comb has narrow, close-fitting teeth; the setae (spines and hairs) are long and numerous; and the body is laterally flattened so the flea can easily move through feathers.

structure, called a sensillium, at the top of the hindbody, which may be used to send and receive ultrasonic songs that are likely used for advertising their location on a host and finding mates.[5]

Fleas are, of course, most famous for jumping. They leap onto hosts using their enlarged hind legs. In addition to the muscular power of their legs, the pleural arch of the thorax contains an elastic protein called resilin that contributes to their famous leaping ability. After launching themselves into space, fleas somersault in the air with their middle pair of legs held out away from their bodies. These legs have grappler-like claws on their ends that can latch onto a passing host. Bird fleas have strong spines on the lower part (tarsus) of each leg that help adult fleas to cling tenaciously to the host;

these tarsal spines are yet another special modification found on bird fleas but not their kin who live on mammals. Many studies indicate that once they detect vibrations, carbon dioxide, and heat, fleas leap randomly and gamble that they will contact a host.

Once on their host, fleas move down onto the skin's surface and begin to feed. Their name Siphonaptera is a good one—their mouthparts form straws that are inserted through the skin until a blood vessel is pierced. Each straw is made up of three parts that come together neatly: two maxillary lacinia (shaped like sword blades), which have rows of backward-facing teeth that cut the skin and help the lacinia to stay inserted, and a stiletto-like epipharynx that is squeezed inside the maxillary tube. The inside diameter of the feeding tube varies in different species, but it is larger for fleas that feed on birds with larger red blood cells, a factor that helps to determine the range of suitable hosts for a species of flea. The larger the diameter of the feeding tube, the faster the flea can feed, and thus, the less chance there is for the host to discover and kill it. However, there is a trade-off between feeding efficiency and how easily (and painlessly) the skin can be penetrated. Therefore, most fleas converge on a maxillary diameter of ten to twelve microns (one micron is equal to 1/1,000 of a millimetre).[6]

Fleas taste their victim's skin using two labial palps. Once a feeding site has been selected and the straw has been inserted, the flea spits saliva (which contains anti-clotting agents) into the wound. Unfortunately, this sometimes allows microparasites, such as virus particles and bacteria (including *Yersinia pestis*—the bacterium that caused the bubonic plague), to be injected into the host. The salivary chemicals often induce a local immunological reaction in their host that leads to a red, swollen, itchy welt where the flea has fed.

Starved fleas feed voraciously, sometimes for hours at a stretch. Males tend to feed more frequently than females but take smaller meals. The blood passes from the mouth back down the digestive system to a small chamber lined with spines that break apart the ingested blood cells. Adult fleas may appear to be wasteful feeders, taking in much more blood than they need; in fact, the undigested blood sprays out of their anus, often into a bird's nest, where it serves as food for flea larvae.

Sexing fleas is easy—the female is larger and the end of her body comes to a blunt taper, whereas the male is smaller with a jaunty, upturned end. Female genitalia include an organ called the spermatheca that is used to store sperm. The spermatheca allows females to fertilize eggs as they ripen, sometimes months after copulation. This structure can easily be seen in specimens under a microscope. It is a bean-shaped organ with a thick neck; its shape, size, and whether or not it is paired are helpful in identifying the species of flea in which it is found.

The genitalia of male fleas, also visible microscopically, include the most complex copulatory organs known in the animal kingdom. Parts on the end of the organ are used for grasping skittish females. The penis itself includes rods that are used to direct it into the female's genital aperture. Its mechanism has long been a source of wonder and bewilderment to zoologists.

Fleas undergo a long and intricate courtship—which probably includes singing—before copulation. The female climbs on top of the smaller male, who erects his antennae to stroke or simply hold on to the bottom of her head. Then the male's complex organ, with its guiding rods, is everted to curl upward and around to the genital opening of the female. Males often die shortly after mating. The elaborate courtship and complex copulatory organs of fleas act as reproductive isolating mechanisms, ensuring that cross-fertilization between different species is rare.

Fleas are anautogenous parasites—this means that female fleas must have a blood meal to produce eggs. In fact, some fleas require a blood meal before they will copulate. The fecundity of the female is affected by the source of blood; fleas will feed on birds (and sometimes mammals) that are not their normal hosts, but when they do, they produce fewer eggs. Hormones in the host's blood are frequently used as signals that fleas can use to coordinate their reproduction with the host's—a fresh batch of nestlings ensures that there will be lots of new blood ready for the next generation of fleas.

Fleas normally produce fewer than twenty relatively large eggs each day, but fleas with long lifespans (of a hundred days or more) may be able to produce several hundred eggs. The creamy white eggs are large enough to be seen with the naked eye. The eggs are smooth, so that when laid, they will roll off the host into its nest.

The survival of eggs and the time it takes for larvae to develop depend heavily on the environment in the nest, particularly its humidity and temperature, both of which tend to be higher when nestlings are present. Flea larvae are maggot-like, with no limbs, and are covered in large bristles. They tend to move away from light and burrow down into the nest material, where they feed on organic debris and fecal blood that has been provided by adult fleas. Larvae moult three times, usually over a period of several weeks, although the larval span can last for several months if the host is absent from its nest for long periods of time. During the pupal stage, the flea rests in a silken cocoon for several days. Like the larva, the pupa is very sensitive to temperature and humidity, and can, for many bird-infesting species, remain in a resting stage for several months, while it awaits the return of the host to the nest. Some species show remarkable environmental adaptations. For example, the larvae and pupae of an Antarctic bird flea (*Glaciopsyllus antarcticus*) that infests colonial, cliff-nesting fulmars (*Fulmarus glacialoides*) can survive freezing temperatures for nine long months.[7] When stimulated by vibrations and chemicals, hungry and not-fully-mature adult fleas emerge from the pupae and jump onto a host to feed.

THE EVOLUTIONARY DANCE BETWEEN HOSTS AND THEIR FLEAS

Because of their life cycle, bird fleas spend most of their lives off their host, burrowed in the nest. Hence, they do not have the same intimate, life-long association with their hosts that feather lice do, so we might not expect to find the same degree of specialization, including the tight co-speciation and co-adaptation patterns, that we saw for lice in Chapter 2. Fleas must also contend with features of the external environment (long periods of cold or dry conditions with no food) that lice don't have to worry about. However, the fact that fleas are obligate blood feeders means that they have a very specialized relationship with birds, a relationship that engages their host's physiological and immunological systems. Aside from the distinct ways in which the bodies of fleas have adapted to the parasitic way of life on a bird, are there other adaptations they have had to develop? And what about their hosts? Have fleas been significant enough players in the world of birds for their hosts to have evolved special defensive tactics?

Fleas are relatively recent parasites of birds, having first sucked blood from burrow-dwelling terrestrial and nest-building arboreal mammals. It is estimated that fleas made this jump from mammal to bird hosts about sixteen times in their evolutionary history.[8] Today, five different families of fleas are found in the flying zoo, and one family (the Ceratophyllidae) includes almost two-thirds of all bird-parasitic flea species.

Despite their athletic reputations, it turns out that not all fleas are good jumpers. Bird fleas come in two types, each type having quite different ecological characteristics from the other. Nomadic species target a wide range of hosts and can live in the nests of birds that may not return to the same nest the following year. When their host is home, they feed frequently, digest their blood meal quickly, and rear several broods of young. They tend to spend the winter in the cocoon, and when a host happens by, they are quickly activated and jump for their lives. They may be somewhat mobile and can be found some distance away from the nest where they hatched. Their bodies are usually rigid and dark (to delay desiccation), they have large eyes, and they are excellent jumpers. It is not uncommon to find adults busily feeding on their host bird. Examples of nomads include the hen flea, *Ceratophyllus gallinae*, as well as *C. garei*, *C. styx*, and *Dasypsyllus gallinulae*.

In contrast, settled fleas are not as adept at jumping as nomadic fleas. They are much more host-specific and prefer birds that faithfully return to the same nest each year. They produce fewer eggs and normally have a single brood. Adults can remain inactive, languishing without food in the nest for months (they have large, fat bodies that store energy). Their vision is not as good and they are not as athletically gifted as nomadic fleas, but they have more flexible bodies that can expand to take in a large blood meal. Settled species are typically found in the nest rather than on a host. Examples of settled fleas include species of *Actenopsylla*, *Glaciopsyllus*, *Ornithopsylla*, *Ceratophyllus farreni*, and *C. hirundinis*. The many distinctions between these two ecotypes of fleas (nomads and settlers) mirrors the division that we saw between ischnoceran and amblyceran feather lice in Chapter 2.

For both nomadic and settled fleas, the characteristics of their hosts' nests are crucial. Even fleas such as the nomad *C. gallinae* (the hen flea), which have been found on more than eighty different hosts occupying many different types of nests—including holes in trees, dry ground nests, and in burrows—will not infest waterfowl, in all likelihood because their nests are too damp. In an analysis of the host preferences of fleas in Britain, Adrian Marshall showed that there were clear patterns of host selection based on where their nests were situated, some of which are summarized in Table 3.1.

The spatial distribution of hosts' nests is another feature that is associated with the biological attributes of bird fleas, as explored by Frederic Tripet and his colleagues.[9] Not surprisingly, territorial birds with nests that are evenly separated in space rather than clustered together (for example, red-winged blackbirds or robins) are usually infested with nomadic species of fleas. However, we can venture some additional, more specific predictions about fleas based on the consequences of occupying nests that are segregated in space. In order to achieve the genetic diversity that's needed for a healthy flea population, bird fleas must move among nests. When the hosts' nests are segregated in space rather than clustered together, environmental conditions can differ quite a bit from nest to nest. These factors should favour high mobility among bird fleas that attack hosts in segregated nests, as well as the ability to tolerate broad differences in humidity, temperature, and nest texture. As a result, these fleas should develop the capacity to parasitize a number of different bird species.

On the other hand, birds that nest in colonies (like swallows and martins) are infested with settled species of fleas. We would expect this trend to be especially strong for birds that nest in colonies consisting of a single species in geographically isolated locations such as islands. Such a nesting pattern should result in reduced genetic diversity among flea populations, reduced flea mobility, increased specialization, and perhaps even flea speciation.

Tripet and his colleagues tested these predictions in 58 species of bird fleas belonging to the family Ceratophyllidae, infesting 117 genera of birds. They found that colonial bird fleas had reduced mobility (as determined by the height of the pleural arch of the thorax, where the important "jump protein," resilin, and the main jump muscles are located). They also had

TABLE 3.1. Relationship between flea ecotypes (N = nomad; S = settled) and the nest types of their hosts

Flea	Ecotype	No. of hosts	Nest type	
			Dry ground	Wet ground
Callopsylla waterstoni	S	1		
Frontopsylla laeta	S	1		
Ornithopsylla laetitiae	S	2		
Ceratophyllus columbae	S	1		
C. farreni	S	1		
C. fionnus	S	1		
C. fringillae	S	2		
C. hirundinis	S	1		
C. rossittensis	S	1		
C. rusticus	S	1		
C. styx	S	1		
Dasypsyllus gallinulae	N	68		X
Ceratophyllus borealis	N	28		
C. gallinae	N	75	X	
C. garei	N	59		
C. vagabundus	N	9		

Adapted from: Marshall, A.G. (1981). *The ecology of ectoparasitic insects.* New York: Academic Press.

narrower host and geographic ranges. Birds with aggregated nests were infested with more species of specialist fleas than hosts with solitary nests. This study demonstrated that the spatial distribution of birds' nests has a great effect on the evolution of their fleas.

Have fleas in turn had any effect on the evolution of birds? Is there any evidence for co-evolution? As we saw with lice in Chapter 2, hosts do not sit quietly by and let parasites have their way with them. Fleas suck blood and they inject chemicals—and sometimes bacteria and virus particles—into

Tree branches	Hole in tree	Hole in cliff/bank	Hole in soil near sea	Ledges near sea	Under overhangs
					X
					X
			X		
					X
					X
			X		
	X				
					X
X					
					X
		X			
	X			X	
				X	
X	X				X
X					
	X				X

birds. Even though fleas spend more time than lice off their host and in the nest, they can still cause serious harm. For example, Heinz Richner and his colleagues discovered that hen fleas (*Ceratophyllus gallinae*) severely affected the reproductive success of great tits (*Parus major*).[10] In experiments that controlled flea infestations in the nests of great tits, the presence of fleas increased the mortality of chicks more than three-fold, significantly lowered fledgling success, led to smaller body mass in chicks, and caused birds to grow to smaller adult size. Not surprisingly, hematocrit levels—a measure of blood cell volume that indicates anemia—were much lower in birds from flea-infested broods, and chicks from these broods were also in much worse nutritional condition. Fleas clearly affected the number and quality of great tit offspring.

Another study of this same bird–flea pair showed that, compared with their unmolested peers, nestlings in infested nests more than doubled their begging behaviours in attempts to get more food—perhaps to mitigate the nutrition-depleting effects of fleas.[11] This resulted in greater food competition among siblings and caused male parents to increase the frequency of feeding trips. However, females did not adjust their provisioning rate, suggesting that they may have made a calculated reproductive decision to save energy for future broods rather than attempt to help their current flea-bitten young. Females did, however, sacrifice a third of their nightly sleep time, which they devoted instead to nest sanitation.[12] These studies confirm that fleas can have major effects on the behaviours of birds.

These two studies looked at the influence of fleas in a single breeding attempt, but a study conducted over four years—the entire lifespan of the great tit—showed that hen fleas have a great impact on the lifetime reproductive success of their hosts.[13] Female great tits were unable to compensate for the chick mortality caused by hen fleas—the fleas were a constant evolutionary force that affected the total number of offspring that a mother could produce in her lifetime.

In addition to affecting their host directly, hen fleas also engineer the environment of their bird's nest to suit themselves, which may have consequences for other parasitic occupants of the nest. Philipp Heeb and his colleagues found that hen fleas living in the tree-cavity nests of great tits affected another of the bird's ectoparasites, the blowfly (Protocalliphora).[14] Blowflies preferred to colonize nests with relatively low humidity, but the presence of fleas caused nest humidity to increase. Although the mechanism by which hen fleas accomplished this was not clear, Heeb and his colleagues speculated that fleas caused the metabolism of tit nestlings to increase (perhaps by forcing them to beg for food more energetically), thereby increasing evapotranspiration from their bodies and resulting in an overall humidity increase. Another possibility was that the wasteful feeding of adult fleas, which released a fecal-blood mixture that was 80% water, increased the humidity of the nests. Regardless of the underlying mechanism, the increased nest humidity had two consequences: it improved the larval and pupal development of the fleas, and it reduced the number of blowfly pupae. Thus, the fleas managed to not only engineer the nest environment to help themselves, but also to harm a blood-meal competitor! Since blowflies can harm great tits by causing their young to be small,[15] in a perverse way, these fleas may have actually been helpful to their hosts.

Even the threat of flea infestation seems to be enough to affect birds, as discovered by Erin O'Brien and Russell Dawson when they placed fleas on the outside of nest boxes used by tree swallows (*Tachycineta bicolor*).[16] Although the birds were never directly exposed to flea bites, when they perceived the risk of infestation, they produced smaller clutches and smaller broods at hatching. Christopher du Feu found that, just by adding artificial fleas (small pieces of black wire insulation) near the entrances of nest boxes, great tits could be induced to make fewer visits and spend less time there—the birds were deterred by the fake parasites.[17]

THE HOST'S DEFENSES: BEHAVIOURS AND BIOLOGY

Birds have options other than giving in to fleas or delaying reproduction in response to the threats posed by these parasites. Behaviourally, birds can try to avoid or abandon nests that have fleas. Birds that use the same nests year after year may employ more creative methods—for instance, adding fresh plant material such as wild carrots or fleabane, which include aromatic compounds, may help to repel their ectoparasites.

Birders who provide artificial nest boxes may attract some birds, but discourage others, by removing old nest material. For example, Eurasian tree sparrows (*Paser monantus*) prefer nest boxes that they have previously defecated in—possibly because the ammonia from the waste kills ectoparasites.[18] Usually, however, ectoparasite populations are denser in previously used nests and birds often avoid them.

The strangest behavioural method for dealing with fleas has been reported for eastern screech owls (*Otus asio*), which use the same nest over a period of several years. Owls capture and transport live blind snakes (*Leptotyphlops dulcis*) to their nests. The snakes burrow into the nest material and feed on fly larvae, beetles, and other arthropods, including fleas. One study showed that up to 90% of the blind snakes found in screech owl nests had scars, which indicated that they were not simply crawling up into trees, but had been captured and transported by the owls. Owl nestlings in nests with snakes grew faster and had lower mortality rates.[19]

Female birds that have been bitten by fleas react immunologically to chemicals in the flea saliva. These birds produce antibodies (immunoglobulins), including ones that are passed on to their eggs. In one study, great tits were either exposed or not exposed to hen fleas after laying the first egg in a clutch. Subsequent eggs were collected, and it was found that the concentration of immunoglobulins increased from the first to the eighth egg in response to the fleas.[20] Another experiment included cross-fostering the young after hatching, and again it was found that antibodies were passed on to eggs laid by mothers that had been exposed to fleas. The young that hatched from these eggs grew faster than the young from eggs without antibodies, proving the effectiveness of the maternal response at egg laying.

The ability of birds to fine-tune the composition of their eggs may be remarkably specific. Erik Postma and his colleagues found that great tits managed to not only increase immunoglobulin antibodies, but also decrease the yolk androgens in eggs—steroids that suppress immunity—depending on their exposure to hen fleas.[21] This exciting research stimulates many questions: How is this feat accomplished? Is the ability genetically inherited? How do the fleas cope in response to these defenses?

In addition to stimulating the production of antibodies, fleas induce birds to employ the second branch of the immune system, the cell-mediated branch. This involves recruiting thymus-derived white blood cells called T-lymphocytes, or simply T cells, which recognize and attack foreign invaders in the body such as viruses, bacteria, or parasites. Experimental studies of domesticated fowl have shown that increased T-cell responses in these birds are associated with fewer fleas and better survival rates. Anders

Moller and his colleagues examined 42 species of birds and 48 species of fleas (all members of the family Ceratophyllidae), and found some interesting correlations involving T-cell responses in birds, the degree to which birds nested in tightly packed colonies, and the numbers and specializations of the fleas. Birds with strong immune responses were mainly colonial nesters. These birds were infested by more species of fleas, and the fleas themselves tended to be specialists that infested only a limited number of hosts. In fact, the number of fleas per host was a good indication of the strength of the bird's immune response. Moller and his colleagues concluded that host defenses and flea specialization have co-evolved, with strong T-cell responses in birds nudging fleas toward greater specialization.[22]

Birds with weak immune responses can host fleas that are specialists (infesting only a few types of host) or generalists (with a wide host spectrum, like the hen flea C. gallinae), but birds with strong immune responses host only specialist fleas that have evolved the ability to overcome the host's immune responses. As parasites become specialists, they often become more virulent; nestlings must therefore invest more energy into defending themselves, at the cost of slower growth and smaller size. It is not clear why colonial hosts have stronger immune responses, but one possible explanation is that colonial nesters are especially vulnerable to flea infestations, given the ease with which fleas can move from nest to nest. It does seem clear that intense selection pressure involving the immune system by birds against fleas has promoted flea specialization and likely flea speciation as well.

Even the hen flea, a prime example of a generalist flea that has been reported from 80 species of birds and 15 mammals, humans among them, appears to be subject to co-evolutionary pressures.[23] Hen fleas seem to indiscriminately attack any forest bird that nests in holes, and have even been reported from crows, Corvus brachyrhynchos, in tree-crown nests, and in nests located in shrubs close to the ground. (According to Miriam Rothschild and Theresa Clay, hen fleas successfully invaded North America when they escaped from poultry farms.[24])

Because hen fleas attack hosts in widely spaced nests that often have quite different microclimates, they are very tolerant of a variety of nest conditions and move around a lot. This nomadic lifestyle would seem to promote gene flow among hen flea sub-populations, and should therefore limit the opportunities for hen fleas to become specialists. Moreover, hen fleas must be flexible enough to inhabit very different host species, even if the fleas stay in the same nest. This is because nests made in holes, common habitats of the hen flea, are often occupied by different species. Hole-nesting birds frequently use naturally occurring holes, rather than excavating their own; the holes often start by being small and are therefore only accessible to small birds, but later enlarge due to natural rotting or the actions of birds such as woodpeckers, allowing larger hosts to occupy them.

In spite of all these factors that should preclude hen fleas from becoming specialists, there is evidence for a close pairing between hen fleas and their hosts: two species of hole-nesting parids, blue tits (*Parus caeruleus*) and great tits (*P. major*), are more frequently infested by hen fleas than any of the numerous other birds that may carry them, with each host of these two species typically afflicted by a greater number of hen fleas than are found on other birds. Thus, even for this poster child of a generalist flea, the stage is set for a co-evolutionary relationship to develop.

Fleas are relative newcomers in the flying zoo. Nevertheless, they still show remarkable facets to their biological interactions with birds. The co-evolutionary dance between these blood feeders and their avian hosts includes unique behavioural, physiological, and immunological adaptations by birds and even more unbelievable adaptations by fleas, including the ability to change their host's nest environment in order to suit themselves. After considering these relationships, I can only agree with the artist who was hired by the microscopist Antonie van Leeuwenhoek to draw images of a flea: "Dear God, what wonders there are in so small a creature!"[25]

4

T O U G H
T I C K S

A HARSH LOCALE

Located in the South Indian Ocean, in the middle of an expanse of water
that stretches from Africa to Australia to Antarctica, is the most remote
place on Earth—the Crozet Archipelago. The six small volcanic islands
(amounting to about 350 km²) that make up this sub-Antarctic archipelago
are perpetually cold, wet, and very windy: the average temperature is 5°C,
rain falls on more than 300 days of the year with an annual accumulation
of more than 2,400 mm, and on 100 of these days, wind speeds reach more
than 100 km per hour.

Until the 1960s, the islands were uninhabited except for occasional
survivors of frequent shipwrecks. In 1821, survivors of the British sealer
Princess of Wales spent two years on the Crozet Islands, and in 1887, survivors of the French vessel *Tamaris* desperately tied a note to the leg of
a giant petrel (*Macronectes giganteus*), which—miraculously—was found

seven months later in Fremantle, Australia. Unfortunately, the crew was never found.

In 1938, the French protected the archipelago as a national park and in 1961 established a permanent research station there—for a very good reason. This bleak group of small islands has a greater population density of birds than any other place in the world. It is estimated that about 25 million birds come to the Crozet Islands to breed.

These dormant volcanic islands have been glaciated, uplifted, and eroded into mountainous terrain slashed by sheer cliffs. The rough, basaltic rock beaches are constantly pounded by breakers slammed onto the shore by the Roaring Forties winds that are so prevalent in the southern hemisphere between the latitudes of 40 and 50 degrees. The meager vegetation is a sub-Antarctic tundra, dominated by stubborn grasses, herbs, shrubs, mosses, and lichens. There are no trees—they cannot survive the relentless winds.

Around the islands, however, the South Equatorial and Western Australian currents bring in rich loads of nutrients, which support sumptuous populations of squid and fish. These in turn feed more than 60 species of seabirds, 7 species of cetaceans, and 3 species of seals. Within this biodiverse environment is the largest colony of king penguins (*Aptenodytes patagonicus*) in the world, with some 300,000 breeding pairs. There are also over 2 million pairs of macaroni penguins (*Eudyptes chrysolophus*) and untold numbers of *Ixodes uriae*, a blood-sucking tick. As amazing as the prosperity of Crozet Island penguins is, the adaptations and adjustments the ticks have made to survive there are even more astounding—these ticks are tough!

Unlike the other ectoparasites we have already met, ticks are not insects. They are arthropods that belong to the same group as spiders and scorpions, namely the chelicerates. Ticks have bodies that include two distinct regions with no distinct head. They lack antennae and have eight legs; the group to which they belong is named after the first pair of appendages (chelicerae) that are used for feeding. Ticks feed on pooled blood after the chelicerae have ripped and torn their host's skin.[1] A structure on the underside of the head, called the hypostome (*hypo*—"below"; *stoma*—"mouth"), is armed with rows of recurved denticles that assist in anchoring the tick's mouthparts into the host. Some ticks also secrete chemical cements to further anchor them.

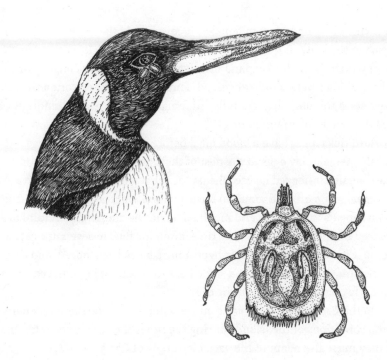

FIGURE 4.1 A king penguin (*Aptenodytes patagonicus*) attacked by two engorging female ticks (*Ixodes uriae*). The enlarged view is of a male tick. Ixodid ticks are called hard ticks—note the shield on the back of the male that almost covers his body and the mouthparts (pedipalps, chelicerae, and hypostome) that project forward.

Like fleas, ticks inject salivary secretions, including a cocktail of more than five hundred different proteins, into their host; these serve to stop blood from clotting, to counteract pain and itch, interfere with immune responses, and so on.[2] Again, as with fleas, this often means that bacteria and virus particles are injected into the host—ticks are notorious for transmitting a great number of infectious diseases.

On the Crozet Islands, *Ixodes uriae* feed on penguins. Ixodid ticks are called hard ticks because they have a hardened (or sclerotized) shield that covers the top side of their bodies, giving these ticks the appearance of miniature turtles with protective carapaces. In male hard ticks, this shield covers the whole body, but in females, it covers only the leg-bearing, anterior part of the body. The female's outside surface, or cuticle, is extensively folded, which allows it to expand massively when the tick is engorging on her host's blood. When fully engorged, female hard ticks are often the size of a person's thumbnail, and a fully fed female tick can weigh hundreds of times the weight of an unfed tick.[3]

Hard ticks must have a blood meal before they can produce eggs. Typically, females lay eggs in the nest of their host; these eggs hatch to release small, six-legged larvae. The larvae must get on to a host, take a blood meal, drop off, and moult to form a larger, eight-legged nymph. The nymph repeats the process and moults to form the adult stage. Adult hard ticks feed on their hosts for a long time, often for four to seven days at a time, so it is important that their hypostomes be well-anchored, and that the ticks attach to their host at a site where they cannot be removed easily—frequently on the head around the eyes.

On the Crozet Islands, *I. uriae* must contend with the stressful environmental conditions of the islands during the times they are off penguin hosts, and they must also adapt to the variable supply of hosts.[4]

WHEN HOME IS A PENGUIN

King penguins have a very unusual breeding system. The birds form dense colonies located among tussocks on gently sloping beaches, usually near the sea. One colony on the Crozet Islands has 200,000 breeding pairs that are squeezed together at a density of more than two pairs per square metre. The birds do not make a nest but balance their single egg—and later, their chick—on their feet. In this harsh environment, it takes more than a year from the time the egg is laid to the time the chick is reared. The parents take turns incubating the egg in shifts of six to eighteen days. During the winter, the chicks have to survive prolonged fasts, sometimes for more than five months, before they are fledged at about one year of age.[5]

In the Crozet archipelago, the ticks also infest macaroni penguins (*Eudyptes chrysolophus*). Like king penguins, these smaller penguins nest on rocky slopes among tussocks, but unlike the king penguins, they often make a nest of pebbles, mud, and sand that they scrape from the beach. Macaroni penguins lay two eggs, but the first is less than half as large as the second and seldom survives. Both parents incubate eggs, taking turns foraging for about twelve-day periods. Incubation takes about thirty-five days, and young are fledged after about sixty days. Unlike king penguin colonies, in which some adults and young are continuously present, macaronis are migratory and are therefore absent from the breeding colony for about seven months, from March or April until they return in October or November.

The developmental pattern of *I. uriae* is typical of hard ticks, following a progression from egg to larva to nymph to adult. The eggs are laid on the ground, and after hatching, each active stage seeks a host, takes a blood meal, and then drops off to develop on the ground (this represents a so-called three-stage life cycle). Periods of time when adult penguins are foraging at sea are troublesome times for the ticks—blood meals are necessary for female ticks to lay eggs and for larvae and nymphs to moult. Females need six to seven uninterrupted days on their host to fully engorge. This is a dangerous time because the penguins may preen using their feet or beaks; however, if a tick feeds on an adult king penguin during a period of incubation or fledgling, the bird cannot use its feet because this is where the egg is incubated and the chick is brooded. To further minimize the dangers of preening, ticks feed on the necks and heads of adult penguins, especially around the eyes—sites that are hard to reach.

Yves Frenot and his colleagues have reported adult penguins infested with several hundred ticks per bird.[6] However, when the host leaves the colony to feed, unimpeded by egg or chick, it is free to scratch with its feet, and once at sea, the ticks will be submerged in cold sea water for more than two weeks. Not surprisingly, after their long swim in salty sea currents, penguins returning to the colony to report for incubation duty are tick-free.

As you can see, for ticks, timing is everything. Male *I. uriae* don't feed but wait patiently for a female. Sometimes, mating will take place as soon as an adult female tick has completed her moult. Copulation takes twelve or more hours and has been observed both on and off the host. Occasionally, several males can be seen surrounding a copulating pair, waiting their turn to mate, and analyses of batches of eggs indicate that different males may father eggs in the same batch.[7] This is good for ticks because it promotes genetic diversity among the offspring, making the tick population more resilient in the face of an extreme physical environment and the development of immunological and physiological resistance in their hosts. The flexible mating system of *I. uriae*, which allows copulation to occur before or after engorgement, helps to ensure that females will be able to produce eggs even if they are colonizing a new site.

The ticks that use macaroni penguins as their hosts face some specific challenges. The penguin parents spell each other off during incubation and brooding, taking turns to forage at sea—this shortens the length of time during which a tick can occupy a host without being dunked in the sea. Furthermore, the birds are migratory, leaving the colony and separating themselves from the ticks for about seven months of the year. In addition, macaroni penguins do not incubate their eggs on their feet, so they can scratch and remove irritating ectoparasites. After arriving at the colony, birds moult their seagoing plumage and are vulnerable to infestation. Paired macaroni penguins allopreen—that is, preen each other—as part of their courtship behaviour. This has been shown to be an effective way of getting rid of ticks; for instance, one study conducted in March at the Crozet Islands found ticks on 19 out of 20 unpaired penguins but on only 8 out of 20 paired birds.[8]

In order for *I. uriae* to be able to use two hosts with different patterns of attendance on the islands, they have had to be adaptable in their life cycle. Surprisingly, the ticks infesting king penguins have fewer feeding opportunities (once a year) than those using macaroni penguins (twice a year). This is because adult king penguins brood their young for exceptionally long periods of time, so when the non-brooding parent goes to sea to feed, it takes a long time before returning to the brooding parent and the chick. Macaroni penguins spend about the same amount of time as king penguins on land for the purpose of courtship, incubating, and brooding young, but they also return to their colony for an additional 25 continuous days to moult. This affords ticks two opportunities to engorge, and is reflected in the length of time needed to complete their life cycle: ticks at the king penguin colony need three years to complete their life cycle, while those using macaroni penguins complete their life cycle in two.

Once engorged, female ticks lay their eggs in crevices and under stones. The eggs may take up to 190 days to hatch depending on temperature. The larvae and nymphs both need at least 6 continuous days to take blood and then another 100 days each to moult—again, with some variability due to temperature.

During heavy rains (which are common in the Crozet Islands), local flooding can wipe out the ticks by washing them away—even though they can survive being submerged for more than 200 days.[9] Adult king penguins moult in very wet habitats that are unfavourable for ticks. Curiously, king penguin chicks are rarely used as hosts by the ticks—although studies of other species (for example of royal penguins, *Eudyptes schlegeli*, at Macquarie Island) have shown that chicks can be heavily infested. Yves Frenot and his colleagues speculate that the ticks' avoidance of king penguin chicks is due to the thickness or some other property of their down, or perhaps can be explained by the reduced activity of the ticks in winter.[10] Another possibility is that ticks cannot distinguish between adult king penguins and chicks, and because infesting adult birds at that time would result in certain death for the ticks when the birds go to sea, they have evolved to avoid feeding on any king penguin during this dangerous time. Whatever the reason, sparing the chicks is a good strategy in the long term; regular infestation of penguin chicks could be a major cause of mortality that, over time, would reduce the host population.

Paleontologists suggest that king penguins evolved about 3 million years ago, in the late Pliocene. Hard ticks probably evolved at least 40 million years ago,[11] so the relationship between these partners has resulted in adaptations and co-adaptations. The hardy *I. uriae* can survive—and even thrive—in an environment where it must contend with host defenses and confront freezing, drowning, scouring by wind and water, and starvation for months at a time.

The amazing adaptability of *I. uriae* has allowed it to infest colonial seabirds in both hemispheres around the world. These ticks can be found in colonies of waved albatross (*Phoebastria irrorata*) and various species of boobies (*Sula*) in the hot, equatorial desert islands of the Galapagos, in colonies of guillemots (*Cepphus grylle*) and kittiwakes (*Rissa tridactyla*) on the sheer cliffs of the northwest Atlantic, in puffin (*Fratercula cirrhata*) colonies in Alaska, and of course, on the penguins of the Crozet Islands. *I. uriae* is a generalist parasite. It has evolved adaptations to allow it to survive in a wide variety of environments and on a wide variety of hosts, although genetic studies indicate that the northern and southern populations are diverging and perhaps should now be considered separate species.

A DESTRUCTIVE PEST

In a life cycle that may take three years to complete, *I. uriae* may spend as little as three weeks (less than 2% of their lifespan) feeding on their penguin hosts. Can this transient interaction be a significant enough factor for these ticks to be an important part of the penguin's zoo? Absolutely!

Ticks are dangerous parasites. When environmental conditions favour tick reproduction, infestations can be serious, and penguins can suffer terribly. Incubating king penguins must stand immobile, their eggs balanced in the brood patch above their feet, while hundreds or thousands of ticks feed on their bodies. The ticks can consume as much as 10% of the blood of a king penguin (which is second in size only to the emperor penguin, *Aptenodytes forsteri*), causing anemia in their host. Tick saliva may contain paralyzing toxins that impede preening and scratching by the host and increase the host's respiration rate, which accelerates tick feeding.[12] Patches of feathers may be lost as a result of irritation caused by tick activity, possibly leading to hypothermia, especially when the penguins go into the cold sea to feed. *I. uriae* also transmit *Borrelia burgdorferi*, a bacterium that slowly corkscrews its way through tissues, causing rashes, arthritis, and even neurological problems—the telltale symptoms of Lyme disease. Michel Gauthier-Clerc and his colleagues found that 14% of adult king penguins had anti-*Borrelia* antibodies, indicating that they had been exposed to these noxious bacteria.[13]

In other bird populations, infestations of *I. uriae* cause birds (such as brown pelicans, *Pelecanus occidentalis*) to desert their nests, and they directly kill chicks (such as kittiwakes, *Rissa tridactyla*). Among king penguins, even in low-infestation years that were not favourable for tick reproduction, ticks were found to reduce the success of rearing a one-year-old chick from 30% to 18%.[14]

King and macaroni penguins in the Crozet Islands try to reduce damage by preening and allopreening. Perhaps other features of their breeding ecology have evolved, at least in part, in response to the ticks. For instance, the balance between the lengths of time spent incubating versus foraging at sea may help to control tick populations. Also, birds moult in waterlogged areas where ticks may get flushed away. In a king penguin colony, the central area has fewer ticks. There is intense competition among birds for this area. This choice real estate not only offers penguins protection from the unrelenting winds and driving sleet that batter their unlucky peers on the colony's periphery, but for some unknown reason, also reduces their tick loads. Over time, penguins have adapted to ticks by developing physiological and immunological responses that interfere with blood feeding. These innate and adaptive immune responses have been found to impede tick feeding.[15]

SOFT TICKS WITH HARD CONSEQUENCES

Hard ticks attack a variety of hosts, including mammals as well as birds. But another type of tick seems to be more specialized for infesting birds. These are soft ticks, the Argasidiae. Argasid ticks look very different from hard ticks—in fact, in most respects, they are the opposite of hard ticks. First, the part of the head that bears the mouthparts, called the capitulum, does not occur at the front end of the body, so the mouthparts are usually hidden from view. The life cycles of soft ticks involve many nymph stages, all of which take a blood meal, so each tick can attack many hosts during its lifetime. The numerous nymph stages extend the ticks' life cycle, with the result that they need only small blood meals to survive. These characteristics allow these hit-and-run parasites to take advantage of highly mobile migratory hosts that may be gone for months at a time.

When they feed, soft ticks don't engorge. They take many small blood meals and only feed for minutes at a time, usually during the night, so soft ticks must be adept at finding hosts for their frequent meals. Their cuticle is pliable and leathery to allow for fast feeding but probably does not offer the same protection as the hard shield of ixodid ticks. (Of course, hard ticks stay on their host to feed for days and therefore need serious protective armour.) Multiple bouts of fast feeding also help soft ticks avoid most of the immune responses of their hosts. Again, unlike hard ticks, soft ticks will feed first and then digest their blood meal later. The excess water taken in during feeding is secreted from coxal glands, located at the base of the legs (hard ticks secrete this water from their salivary glands instead). Argasids hide in their host's nest or burrow and seem to prefer dry sites. They may incessantly attack the same host individual or family.

Unlike most hard ticks, soft ticks will mate while unfed. Female soft ticks lay smaller batches of eggs in their host's nest (between one and two hundred, whereas hard ticks lay several thousand at a time), which hatch to produce six-legged larvae. Argasids that feed on migratory birds can time their egg laying to coincide with the times at which their host is most likely to be present. This seems to be regulated by sensitivity to the amount of daylight in a 24-hour period.[16] For example, ticks that attack the cattle egret (*Bubulcus ibis*) feed in the late summer and delay egg laying until their migrating hosts return in the spring.[17]

After hatching, larvae feed and stay on their host to moult, producing nymphs that feed repeatedly. One thing that soft ticks have in common with hard ticks is that they are very patient. They can remain in a state of dormancy during hot, dry times for months to several years, waiting for a host upon which they can feed. One species of soft tick, *Ornithodoros lahorensis*, can go without food for eighteen *years*![18] If no bird host is available, starved soft ticks may engage in homovampirism—they will not hesitate to feed on the blood inside another tick.

After feeding, soft ticks detach from their host and shelter nearby to digest blood meals and reduce their chances of desiccation. On African

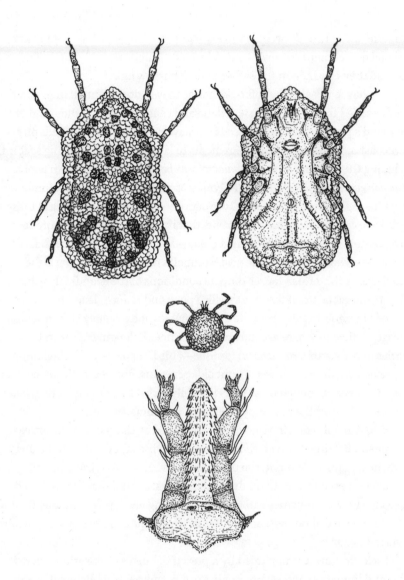

FIGURE 4.2   Dorsal (topside) and ventral (underside) view of an argasid tick, or soft tick, *Ornothodorus spheniscus*. Note that the mouthparts are hidden from view. The middle drawing depicts a larval stage (called a seed tick) with only three pairs of legs. The bottom drawing shows an enlarged view of the tick's mouthparts, including the well-armed hypostome.

penguins (*Sphenicus demersus*), the argasid tick *O. capensis* has what is known as a diel detachment pattern—the ticks feed at night and drop off before dawn so they can hide in the safety of host burrows.[19]

The havoc wreaked by soft ticks has been well documented. In 1862, a fleet of ships sailed from Peru to Easter Island and captured about a thousand of the island's people, who were forced into slavery on the Guano Islands off the west coast of Peru. These islands are home to large colonies of nesting seabirds, including Guanay cormorants (*Phalacrocorax bougainvillii*), Peruvian boobies (*Sula variegata*), brown pelicans (*Pelecanus occidentalis thagus*), and Humboldt penguins (*Spheniscus humboldti*). Over many generations, the wastes of these birds have created thick guano deposits that have been mined for more than three centuries—and are still mined today—as a valuable agricultural fertilizer. (The same nutrient-rich, cold Humboldt current that supports the colonies of seabirds also causes the arid conditions of the islands. This climate inhibits microbial breakdown of the bird poop and reduces leaching of ammonia and ammonium salts from the guano, making it extremely rich in soluble nitrogen.) The birds here are infested by soft ticks belonging to several species in the genus *Ornithodorus* (*ornitho*—"bird"; *doros*—"gift"), including *O. spheniscus*, a tick that infests Humboldt penguins, boobies, pelicans, cormorants, and terns. Although these ticks typically target birds as hosts, they had no trouble feeding off the new supply of hosts, the human slaves.

One can only imagine the torment suffered by the slaves—exposure to the hot sun beating down on these barren islands as they toiled in the dirty, dusty, acrid mines. At night, the nymphs and adults of *O. spheniscus* and *O. amblus* emerged from their lairs under the guano and savagely attacked the workers. Tick densities can reach astounding levels. For example, Conrad Yunker reported about 600,000 *O. denmarki* per square metre around seabird nests in Florida.[20]

Tick bites are accompanied by a piercing, burning sensation. Soon after, inflammation and severe itching occur and painful blisters develop. These blisters take days to recede and are often followed by fever and headaches. After multiple tick bites, gangrene and death can occur. It is little wonder that only fifteen of the kidnapped Easter Islanders survived to be returned to their home three years later.

The ticks also torment the seabirds. David Duffy noticed that young boobies had crusts on their legs where guano had stuck to the open wounds left by the ticks.[21] Humboldt penguins respond by abandoning their eggs, and the other seabirds responsible for producing guano spend more time than usual in flight, away from their breeding and resting grounds. The guano birds also desert their nests. But before leaving their eggs and young to predators such as gulls and vultures, boobies were found to spend five times as long in preening behaviour compared with birds in less-infested colonies, suggesting that nest desertion is a last resort. On Mazorca Island, Duffy reported that about 75% of 385,000 nesting pairs of cormorants, boobies, and pelicans deserted their nests, resulting in a loss of between 250,000 and 750,000 eggs and young! Also, the ticks transmit viruses and bacteria, compounding the damage caused by nest desertion.

Even if the birds completely vacated these breeding grounds for several years, it wouldn't matter—the ticks can survive starvation for years. The only respite from these destructive parasites is provided by tick predators (lizards, spiders, scorpions, and ants), by periodic crashes in the guano bird populations due to El Niño events (when warm currents draw water from the equatorial region, which contains less seabird food), and by the eco-logical disturbance of guano mining, which results in a complete disrup-tion of tick habitat. David Duffy reported that guano extraction has gone on since the mid-1500s; before that, the guano beds harboured burrowing sea-birds, whose burrows may have supported greater densities of tick preda-tors. Nevertheless, if tick populations have historically caused massive nest desertion events, guano birds may have had to avoid the islands for years to starve down tick populations. This would have exacerbated the shortage of suitable nesting sites in the area, leading to selection for denser colonies.[22]

Other soft ticks that belong to the genus *Argas* are even more special-ized for parasitizing birds. One example is the fowl tick, *A. persicus*, which evolved in Central Asia, initially attacking arboreal nesting sparrows and crows. It has managed to switch hosts to infest domestic fowl around the world. Fowl ticks are nocturnal feeders that hide in henhouses during the day. Populations can reach such high numbers that they literally exsanguinate (bleed out) chickens, but there are few reports of the ticks attacking humans. Fowl ticks transmit bacteria that cause relapsing fevers in birds.

One unique soft tick, *A. cucumerius*, also belongs to this genus. This tick infests birds that nest along sheer cliff faces. The ticks feed during the day rather than at night and are very active in locating, chasing down, and feeding upon their hosts. As is the case with almost all ticks, they can survive for months to years without feeding, but when the opportunity arises, they are very aggressive in hunting and attacking their hosts.

Whereas hard ticks are patient, opportunistic parasites that have adapted to their hosts by coordinating their life cycles with the breeding biology of their hosts, soft ticks take matters into their own hands (or chelicerae), seeking out hosts and feeding quickly. The chances for immunological and physiological defenses to develop in birds attacked by hard ticks are better than for those attacked by soft ticks, due to the length of time taken in feeding. Birds try to "defend" themselves from soft ticks by deserting nests and nesting colonies, sometimes for several years, or they may try to choose nest sites in areas that have tick predators (such as ants).

Most ticks are physically attached to their hosts for only a brief part of their lives, but they are still a very significant part of the flying zoo. Their feeding causes direct damage to their hosts, leading to irritation, anemia, toxemia, and even blindness. In addition, they cause indirect damage through bacterial and viral transmission. Such transmission can occur in a variety of ways: ticks can transmit infections to birds through their saliva, their feces, the fluid produced by the coxal glands of soft ticks, or simply by being eaten during preening. Virus particles can be transferred from infected female ticks to their eggs, so that the next generation of ticks is armed with pathogens even before they have fed.

Ticks can transmit infections to humans who inadvertently or intentionally enter their domain—for example, by being forced to work on the Guano Islands of Peru or by doing the work of banding birds. Ticks are nasty, irritating, and dangerous parasites. Nevertheless, we can still marvel at their incredible abilities to endure freezing, heating, drying, drowning, preening and starving—abilities that have allowed them to survive and thrive in the flying zoo for more than forty million years.

MITES

Little Things Mean a Lot

VARIETY IS THE BEST STRATEGY

According to the old proverb, "Little things please little minds." Taken literally, this saying implies that mites are dull, simple, and unworthy of our attention. That idea is completely false. These tiny creatures are some of the most successful, diverse, and weird inhabitants of the flying zoo, and they have surprising and complex relationships with their hosts.

Mites are small ticks—*very* small ticks. Or, more accurately, ticks are really giant mites, because it seems likely that mites evolved long before ticks; mites were probably key members of the soil litter in the coal swamps of the Paleozoic, 400 million years ago.[1] And they are far more numerous and diverse than ticks: whereas there are only about 900 species of ticks associated with all animals, at least 2,500 species of mites associate specifically with birds, and many more species (perhaps close to a million) are found in other habitats. All ticks occur within a single sub-order (Ixodida) in the class Arachnida, while mites not only occur in the same order as ticks (Parasitiformes), but are also found in a second order and in three other sub-orders. Although they are very small animals, they are essential components of the flying zoo.

Mites are micro-chelicerates and thus have the same pincer-like mouthparts as ticks. On average, mites are only about ½ mm to ¾ mm long, just large enough to see with the naked eye. Some, however, are much smaller and can only be detected under a microscope. Their small size is more than compensated for by their abundance, diversity, and the variety of habitats they exploit. In fact, by any measure of biological success—persistence over geological time, abundance, number of species, and the variety of lifestyles and habitats to which they are adapted—mites are winners. They are the commonest and most abundant animals at the flying zoo.

This great biological success is likely a product of their small size. Small organisms tend to have shorter intervals between generations than large organisms, which allows them to reproduce quickly and to develop rapid genetic adaptations. They can also exploit a greater variety of habitats and lifestyles, which together make up an ecological niche.

Mites associated with birds have adapted to various habitats. They can live in nests, as do many species of nidicolous mites, or they can live on (or inside) a bird as commensals—organisms that neither help nor harm their host—or as parasites. Parasitic mites can live inside the bird's body as endoparasites or on the surface as ectoparasites. Those on the surface can live on the skin or on or inside feathers. Those on feathers can choose body feathers or the contour feathers of the wings or tail. Those on wing feathers can live on vanes (the flat parts of the feather on either side of the shaft), or they may live their lives inside a quill. Those on vanes can live on the inner vane or the outer vane, or they can live on the barbs close to the quill or further from the quill on the upper or lower surface. To mites, a host bird contains a multitude of habitats.

In addition to their broad range of home addresses, bird mites hold a variety of different occupations. Mites may live by sucking blood or by eating the keratin of feathers, the spongy medulla inside a quill, or organic detritus from the skin's surface. They may graze on fungi or devour other

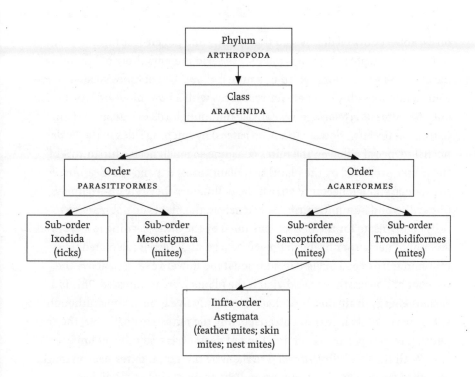

FIGURE 5.1   The relationships between mites and ticks. Note that ticks are derived and more recent than mites.

mites. They may, as do most feather mites, live by eating the waxy and oily secretions of their birds' preen gland, a diet made complete by the dissolved fungal spores and bacteria that are intermingled in these secretions.[2] These various diets give rise to several distinct roles that mites can play in the flying zoo: they may function as parasites that can depress the health of their hosts, as commensals that associate with birds but don't seem to seriously affect them, or even as mutualists that provide some benefit to their host bird. A single species may play any or all of these roles, which may change depending on circumstances.

Incredibly, mites can even be competitors of birds. This relationship was observed in one of the most unexpected interactions between mites and birds. Robert Colwell discovered that nectar- and pollen-feeding flower mites, which are parasites of tropical plants, are serious competitors of humming-birds.[3] Small as a hummingbird is, a mite is thousands of times smaller—how could a mite possibly compete for resources with a hummingbird? As it turns out, the mites, *Proctolaelaps kirmsei*, use hummingbirds to transport them to tropical firebush flowers (*Hamelia patens*) by hitching rides in the birds' nostrils. Once at a flower, the mites consume so much nectar (up to 40% of the nectar produced by the plant) that plants must expend extra resources to produce enough nectar to attract the pollinators they rely on—namely, butterflies and hummingbirds. Nevertheless, the plants do not manage to fully compensate for the nectar consumed by the mites, leading to a reduced amount of nectar per firebush flower. As a result, Colwell calculated that hummingbirds need to visit an additional 160 flowers over and above the number of flowers they would visit if the plants were uninfested. This is a serious energy drain on a bird that weighs only five grams! Thus, although the extra energy costs incurred in transporting the mites are negligible, the competition resulting from the mites' theft of nectar has a significant impact.

With the flexibility afforded to them by their size, mites have managed to embed themselves in an almost endless array of ecological niches.

NEST-DWELLING MITES

Nidicolous mites, or nest mites, are common residents in the cavity nests of birds. Nest mites probably evolved during an intermediate stage between older, litter-inhabiting mites that roamed freely and more recent mites that live their entire lives on birds, with a number of species settling in comfort-ably to this day within the nest-dwelling niche.[4] Most nest mites are para-sitic, with negative effects on their hosts; however, given the diverse habits of mites, their relationship with birds can be complex.

Edward Burtt and his colleagues studied nest mites in nest boxes used by three species of birds in Ohio, eastern bluebirds (*Sialia sialis*), tree swal-lows (*Tachycineta bicolor*), and house wrens (*Troglodytes aedon*).[5] They discov-ered that the nests were inhabited by three types of mites: a parasitic species

(*Dermanyssus hirundinis*), a scavenging mite (*Dermatophagoides evansi*), and a predatory mite (*Cheletomorpha lepidopterorum*). The parasitic species was a blood feeder, emerging from the nest to feed on nestlings. In addition to consuming precious blood, this mite can trigger allergic reactions and transmit diseases. The researchers discovered that all three birds in this study were attacked by this parasite, with the nests of house wrens sheltering, on average, nearly 13,000 blood-feeding mites.

The scavenger mite found in these nests is a common free-living species that feeds on organic debris. Like the parasite, it was observed in the nests of all three bird species but was not as numerous as the parasitic mite— house wrens had the greatest number in their nests, with an average of about 7,000 mites.

As would be expected based on general ecological principles, which dictate that animals at the top of food webs are less numerous than those lower down, the predator *Cheletomorpha* was the least numerous mite and was absent from bluebird nests. Again, house wren nests accommodated the highest number of these predatory mites, with an average of about 2,500 in each nest. *Cheletomorpha* is an ambush predator. It waits patiently in the nest until its prey moves before striking. Among its most common prey is the parasitic mite *Dermanyssus*, which hides out in the lower parts of nests and, not surprisingly, remains still between feeding bouts to escape the attention of *Cheletomorpha*.

The numbers of parasitic and scavenging mites found in the nests were directly related to the number of nestlings residing there, while the number of predators was related to the abundance of their prey, apart from the curious absence of predators in bluebird nests. The researchers attributed this absence to the architecture of the birds' nests. Bluebirds make flat nests that are densely woven. Consequently, the parasitic mites had only a short distance to move to attack nestlings, and the densely woven vegetation provided good cover for them to escape from predators. Bluebird nests also contained little or no fecal matter, so there was less food available for scavenging mites, again resulting in less prey for predator mites.

In contrast, the nests of house wrens proved to be choice dwellings for predator mites. These nests were composed of coarse materials, including twigs and sticks, which may be the primary vehicles by which predatory mites are dispersed. The nest materials were loosely woven, providing little in the way of cover for mites that served as prey. The nests also contained lots of fecal matter as well as flakes of feather sheaths and other organic debris, which attracted scavenger mites. Moreover, the large number of nestlings in house wren nests resulted in an abundance of food for both parasites and scavengers, which in turn ensured a steady food supply for the predator mites. The presence of predator mites, however, did not seem to have much effect on the numbers of scavengers or parasites; when the nests of tree swallows or house wrens were free of predator mites, their numbers of scavengers and parasites were about the same as those in nests that did contain predators.

There was no relationship between mite numbers and the size of the nests of the three birds, so it seems clear that nest architecture, rather than size, plays a crucial role in determining the community of nidicolous mites.

Although no deaths of nestlings could be directly attributed to parasitic mites in this particular study of nidicolous mites, there was circumstantial evidence that *Dermanyssus hirundinus* may have set a limit to the length of time that nestlings stayed in the nest. A related parasite, *D. prognephilus*, was found to cause purple martins (*Progne subis*) to be lighter at fledging, a factor that can affect adult survival. However, other parasitic mites can have drastic effects on their hosts.

Burtt's study, like all good studies in biology, creates far more questions than answers and has implications that go well beyond simple facts about the specific animals being studied. Studies like these can serve as starting points for a much broader understanding of ecological principles. For example, of what importance is the rate or order of colonization by mites in nests? How important are specific predator–prey relationships in determining the abundance of parasites in nests? If we think of bird nests as analogous to islands, do mite populations in bird nests show the same patterns in biodiversity as are predicted by the theory of island biogeography, which shows a relationship between the size of an island and the diversity of animals living on it? These hypotheses and other ideas about community ecology could all be questioned experimentally, using nidicolous mites as model organisms.

Three species of parasitic mites that attack birds are especially infamous: chicken mites, also known as poultry red mites (*Dermanyssus gallinae*), northern fowl mites (*Ornithonyssus sylviarum*), and scaly leg mites, (*Knemidokoptes jamaicensis*). Chicken mites and northern fowl mites, both of which suck blood, tend to live in or near the nests of birds and climb on hosts to feed. Both have been reported to cause anemia as well as reduced fledging time, decreased hatching success, and weight loss in chicks. Chicken mites have also been reported to transmit equine encephalitis virus. The adverse effects caused by chicken mites should not be too surprising when we consider the short period of time they have between generations, which allows them to develop huge populations—up to half a million mites were estimated to occur in one nest alone.[6] Chicken mites are not averse to biting humans, and they can cause dermatitis and severe irritation. They do not have the elaborate attachment structures seen in other ectoparasites and they attack a wide variety of birds, including many passerines, a fact that may indicate a recent adoption of the parasitic lifestyle. They are mobile, hit-and-run parasites. Chicken mites live mainly off their hosts, but crawl on and off to feed repeatedly at night.

Northern fowl mites, however, complete their entire life cycles on birds. They have been called the worst ectoparasites of poultry in the United States, and for good reason.[7] Northern fowl mites have short life cycles of five to twelve days, and both the protonymph (first developmental stage) and adult feed on blood. Their survival varies with environmental temperatures, and they do best at lower temperatures. Populations can reach densities of 25,000 mites on a single bird.

Northern fowl mites are stimulated by vibrations, and they are capable of orienting toward a source of heat, so they are easily transmitted through bird-to-bird contact.[8] Once on a bird, the mites live beneath the feathers in the vent region, about two centimetres away from the skin surface. They migrate down to the skin surface to feed. A related species, the tropical fowl mite (*Ornithonyssus bursa*), has a similar lifestyle but lives and feeds near the head of its host.

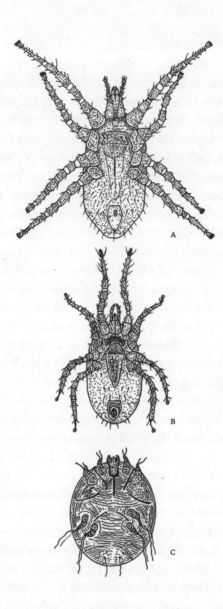

FIGURE 5.2   Three highly pathogenic mites of birds. (A) The northern fowl mite, *Ornithonyssus sylviarum*. (B) The chicken mite, *Dermanyssus gallinae*. (C) The scaly leg mite, *Knemidokoptes jamaicensis*.

Transmission of northern fowl mites occurs vertically, between adults and young, as well as horizontally, from adult to adult. Mites living in nests can also hop onto any birds that move into these nests. This raises some questions: Can the mites be exchanged easily from one species of bird to another? If so, what are the consequences? Are some species affected differently than others by a mite infestation? These questions were addressed in a study by Francisco Valera and colleagues in 2003.[9] The researchers looked at several different ectoparasites that affect European bee-eaters (*Merops apiaster*) and rock sparrows (*Petronia petronia*). These birds live together in mixed breeding colonies in holes excavated in sandy cliffs in southern Spain.

Rock sparrows are freeloaders that use old bee-eater nests. They gain substantial benefits by nesting among bee-eaters, because they do not have to expend the time and energy to dig their own nests. However, one serious drawback of using pre-owned nests is that the rock sparrows become exposed to the previous owners' parasites. But parasites that attack bee-eaters may not have the same effects on rock sparrows. Both the ease of transmission and the consequences of infestation can vary between species, depending on the fit between host and parasite. For example, lice were never exchanged between species, as they are host-specific; bee-eaters were infested with three species of feather lice, whereas rock sparrows hosted a single (different) species.

In contrast, both species of birds were assaulted by a parasitic fly (*Carnus hemapterus*) that attacks nestlings before they are covered by feathers. (A detailed discussion of parasitic flies is coming in Chapter 6.) Flies lay eggs in nests and overwinter as pupae. Adults emerge from pupae, take blood meals from whatever baby birds are in the nest, mate, and fly to a new nest where they lay eggs. Although flies will not hesitate to attack any bird, their life cycle (particularly the emergence of pupae) is timed to coincide with the hatching of bee-eater eggs. Rock sparrows use pre-made bee-eater nests, but they lay their eggs earlier, and their young develop feathers faster, so rock sparrow nestlings are no longer available as a meal by the time the new flies hatch. Thus, the sparrows do not suffer as much from flies as the bee-eaters—which surely does not help deter the freeloading sparrows from relying on the labours of the nest-building bee-eaters.

How do mites fit into this nest-sharing scenario? Blood-feeding fowl mites feed on both species of host, and they can develop large populations—up to five hundred can occur on a single bee-eater nestling. When nesting alone, bee-eaters are not troubled much by mites because they excavate new, clean nests each year. They can successfully fledge their young before mites become too much of a problem. However, when nesting in mixed colonies with rock sparrows, things change. Fowl mites are highly mobile and, after feeding for a generation or two on sparrows and building up their populations, mites can move from pre-owned sparrow nests to new bee-eater nests. Perhaps this is the reason that bee-eaters are aggressive to rock sparrows—bee-eaters get no obvious benefits from nesting together with the sparrows, and they suffer from increased mite and fly attacks as a result of the co-housing situation. The benefits to sparrows, on the other hand, are low-cost housing and diluted parasite loads. The dynamics of these interactions make you wonder if sparrows could be employing parasitic mites as a sneaky weapon of competition, to counter-attack the bee-eaters who try to evict them.

The third parasitic mite in the "axis of evil" trio is the scaly leg mite, *Knemidocoptes jamaicensis*. Scaly leg mites live and reproduce beneath the skin on the legs and feet of passerines and other birds. Infected skin forms warty lesions and generates ribbon-like profusions. Toenails can become so overgrown that the bird cannot perch, and the toes can become necrotic and secondarily infected by bacteria and fungi.

In addition to the pathology clearly seen in individual birds, Steven Latta discovered that scaly leg mites can affect populations of birds, as was the case with palm and prairie warblers (*Dendroica palmarum* and *D. discolor*).[10] Both species overwinter in the Dominican Republic. And both are infested by scaly leg mites, which affect up to 25% of the palm warblers and 7% of the prairie warblers). Latta found that the mites had devastating consequences beyond their usual pathogenic effects: they were also associated with a significant reduction in pectoral flight muscle mass, and most disturbingly, not one infested palm warbler was seen to return to the area after migration.

Latta proposed that both environmental factors and bird behaviour contribute to mite infestations. Palm and prairie warblers that live in drier habitats (conditions in which scaly leg mites survive better) roost communally at night—perhaps because the canopy height and density here are relatively low, so birds gain some protection from predators, according to the "many-eyes" theory. Unfortunately, this habitat is particularly good for spreading scaly leg mites, which live continuously on birds and need close contact for transmission.

As if this weren't bad enough, a drier habitat likely provides less food than wetter areas, which means that the warblers must spend more time foraging and likely suffer from physiological stress and a weaker immune system. Consequently, Steven Latta suggested that birds that chose to live in the thorn-desert habitat in the Dominican Republic were more heavily infested by mites than birds in other habitats, deteriorated in body condition as a result, and failed to return. This created a vicious cycle of habitat affecting bird condition and survival. Latta argued that the mites do not directly limit their host population, but they can mitigate the warblers' ability to compete, which slows population growth. If these mites can so drastically affect individuals and even populations, can they also affect the evolution of birds?

This question was addressed in a study by Craig Benckman and his colleagues.[11] Red crossbills (*Loxia curvirostra*) in western North America have an intimate association with Rocky Mountain lodgepole pine trees (*Pinus contorta*). In fact, it seems likely that the evolutionary histories of the two have become intertwined. Crossbills forage on pine cones, and in response, pines have evolved larger, tougher cones. The birds have in turn responded by evolving larger, deeper bills, with the depth of the bill being a good predictor of the foraging efficiency of the bird. Curiously, however, Benckman found one population of red crossbills in South Hills, Idaho whose bills were not as deep as would be expected for optimal foraging. What was different about these birds?

The researchers found the key to the puzzle when they discovered that the crossbills here were infested with scaly leg mites, and that the mites infested male birds with large bills more heavily than those with smaller bills (although this pattern did not hold for female birds.) As a result, the bill depth of males (but not females) was reduced compared to crossbill populations in other locations. This was because males with large bills were more likely to have scaly leg mites, and males with mites were less likely to survive. However, why were large-billed males especially vulnerable to infestation in the first place? The mites were not choosing birds with larger bills, but bill depth is directly related to body size, and bigger males were getting more infested. Why would this be? Perhaps because transmission of mites requires close bird-to-bird contact. Larger males were more likely to be involved in aggressive encounters, thereby promoting transmission. Also, large males may produce more androgenic hormones, which inhibit the body's immune response.

This situation could give rise to some interesting evolutionary consequences—the mites may be a force that is currently counteracting the pressures of sexual selection, in which females choose larger males as mates. Scaly leg mites could also indirectly affect lodgepole pines, because crossbills with less-than-optimal bill size (for foraging) would have better survival rates, and pines would therefore no longer need to invest as much energy and resources into making tougher cones. In fact, lodgepoles with less investment in cone defense would have more resources available for cone production, and thus would be selected for. Strange as it may seem, scaly leg mites may be driving the evolution of lodgepole pine trees, as well as that of red crossbills, in surprising directions.

MOVING INDOORS

Other parasitic mites have managed to invade deeper into the flying zoo, not as ectoparasites that live on their hosts, but as endoparasites that live *in* their hosts. About 600 species of mites make their homes in the nasal passages of all kinds of birds (except the flightless ratites of the southern hemisphere), where they feed on blood or other tissues. Most of these mites (about 500 species) belong to the family Rhinonyssidae, while the remainder belong to three other families. Rhinonyssid nasal mites can cause lesions and inflammation in the upper respiratory tract, and they occasionally invade the lungs, where they can cause congestion.

Over the course of their evolution, nasal mites have shown a trend toward reduced body size and loss of setae (small, hairlike cuticular projections)—as seen in *Sternostoma* mites found in the nasal passages and lungs of finches. One problem for most endoparasites is that of finding a method for attaching to their host without being evicted. A nasal mite's solution is to use claws or suckers at the ends of its tarsi and embed itself into mucus in the nasal passages.

Nasal mites have become specialized to the point of being very host-specific—they are almost always exclusive to one order, and frequently, to one genus and species, of host.[12] In addition to their attachment structures, their life cycle has become adapted for their parasitic lifestyle—they can live for several months inside their host and can complete their life cycle from egg to adult within a few days. One study even reported that two species of nasal mites can produce eggs with completely developed nymphs (rather than larvae), thus implying that they are viviparous (that is, producing live young), thereby further shortening their life cycle.[13] This would increase the population inside a bird and facilitate transmission to other birds.

No one seems to know for certain how nasal mites are transmitted, but one study hinted that female mites migrate out of the host's nostrils, onto its head plumage, and from there into another bird.[14] Some studies suggest that transmission is from adult birds to their young, but others indicate that it is between adults. Regardless of how infections are transmitted, the method seems to work well; some studies have reported infections in up to 70% of birds in a population (although lower levels of about 15% to 20% are more common). Although they are extremely common, we are left with many mysteries concerning nasal mites and their relationships with birds. They are enigmatic members of the flying zoo and deserve more attention.

There are other endoparasitic mites of birds whose biology is even more mysterious and amazing. Females of mites belonging to the species *Hypodectes propus* lay their eggs in the nests of pigeons and many other birds. After hatching, the deutonymphs (the second developmental stage) penetrate and invade the subcutaneous fatty tissues of their nestling and adult hosts. The mites form white, cyst-like structures, usually near the skeletal muscles, but also in the viscera, near the proventriculus (a bird's glandular stomach), and even around the heart.[15] Incredibly, the deutonymphs (which look like elongated sacs with four pairs of very short legs) have no mouth-parts. After embedding into host tissues, they absorb their food—possibly lipids in these tissues—like living sponges. The parasitic deutonymphs get all the nutrients that the mites will have for their entire lives. After moulting, these mites leave their host and live as adults in the nest, but they do not feed. The adult females have no mouthparts, while males have huge mouthparts that are probably used for reproduction, not feeding. Even though these mites live off the fat of the land in a rather literal way, they seem to have established a benign relationship with their hosts, as they do not appear to be responsible for any major clinical signs or significant pathology.

Another bizarre association between mites and birds was reported by A. Fain and Robert Smiley, who found mites embedded deep in the connective tissues of the trachea and bronchi of great horned owls (*Bubo virginianus*).[16] Considering the frequency of nasal mites and their ability to invade their hosts' lungs, this wasn't too surprising on its own; the surprise came when Fain and Smiley identified the mites as belonging to a new genus and species (*Pneumophagus bubonis*) that belongs to a family of mites called the Cloacaridae. This family includes about twelve species of mites that are all found living in the cloacae (the chamber at the end of the digestive system where eggs and urinary products also collect) of turtles! How could a mite make the transition from the cloaca of a turtle to the respiratory system of an owl? The solution to the puzzle may lie somewhere in the fact that three species of mites have been observed living in loose connective tissues enveloping the muscles and subcutaneous tissues of small mammals such as voles and insectivores, which serve as food for owls. Nevertheless, the mystery remains.

IN THE FEATHERED JUNGLE

The last group of mites that reside in the flying zoo are the ones about which we know the most—and consequently, the ones that stimulate the most interesting questions. Feather mites spend their entire lives within the plumage of their hosts. As a result of their intimate connection with their host, they should demonstrate all the adaptations and specializations we see in feather lice. In fact, we would expect the evolutionary histories of birds and feather mites to be closely intertwined.

There are more than 2,000 species of feather mites, belonging to the sub-order Astigmata. Some species live on the feather surfaces, others on the skin, and still others inside feather quills. More than 24 species of feather mites have been found on just one species of parrot, the green conure (*Aratinga holochlora*).[17]

Feather mites show their trademark diversity, occurring on every type of bird, even penguins—where, for at least fifty years, they were not thought to occur.[18] Some mites feed on the keratin of feather plumes, while others graze on organic debris and sloughed skin. Some suck blood and the fluid in the lymphatic system or eat the soft tissues inside feather quills. Some are predators of other mites. Most, however, live quiet, unassuming lives, feeding on oily secretions from the preen gland, which are fortified with trapped dirt and fungal spores.

As we saw in Chapter 2, the plumage of birds is a rich habitat for chewing lice—for feather mites, which are much smaller than lice, it is a veritable jungle. The jungle analogy was developed by J. Gaud and Warren Atyeo, who compared feather surfaces to tree branches, the skin surface to the ground, feather rachises to tree trunks, and feather follicles to roots.[19]

A jungle habitat gives rise to an abundance and diversity of animal residents. Indeed, with so many potential places to live and different foods to eat, feather mites have diversified and specialized to fill the many ecological niches in this jungle. Many have elaborate body shapes, with asymmetrical appendages or unusual setae, suckers, and spines that give them a highly ornamented appearance. These physical features permit feather mites to wedge themselves between barbules of feathers and to hold on despite the hurricane-force winds of air that flow over the wings during flight.

Researchers Jae Choe and Ke Chung Kim examined the detailed distribution of feather mites on the wings of Alaskan seabirds and found some fascinating relationships between mites and feather structure.[20] Common murres (*Uria aalgae*), thick-billed murres (*U. lomvia*), black-legged kittiwakes (*Rissa tridactyla*), and red-legged kittiwakes (*R. brevirostris*) have

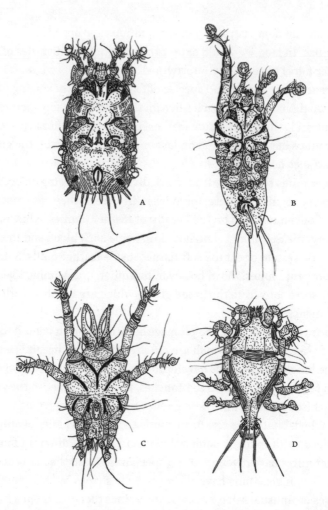

FIGURE 5.3 A gallery of feather mites demonstrating different morphological adaptations. Note the different types of mouthparts, the differences in shapes and sizes of spines, the different ambulacrae (disk or sucker-like appendages used to adhere to feathers or to allow the mites to skate through feather barbs), and the different body shapes, including useful asymmetry. All diagrams show the ventral (bottom) surface of males. (A) *Pelecymerus tetragonus* from a magpie goose, *Anseranus semipalmata*. (B) *Michaelia amplosinus* from a pelagic cormorant, *Phalacrocorax pelagicus*. (C) *Bdellorhynchus polymorphus* from a northern shoveler duck, *Spatula clypeata*. (D) *Hemicalcealges margaropygus* from Ross's turaco, *Musophaga rossae*.

consistent and predictable ectoparasites, including 3 species of chewing lice, 2 species of ticks, and 5 to 9 species of mites on each type of host. These birds also harbour two kinds of feather mites (*Alloptes* and *Laronyssus martini*) that live on the surfaces of wing feathers and a quill mite. Choe and Kim found that *Alloptes* and *Laronyssus* attached themselves on barbs far from the quill, on the bottom surfaces of trailing vanes, where they were protected from the greatest air flow during flight, and where they were also shielded from preening. On the wings, most mites were found on the inner primary feathers (the main flight feathers along the outer edge of the wingtip) and outer secondary feathers (along the back edge of the wing close to the body), and most lived on the mid-sections of the wings rather than on the inner "arm" regions, or the outer "hand" regions.

Choe and Kim suggested that the middle wing section was the optimal habitat for mites, because feathers are attached at the wrist joint and overlap when wings are folded. This provides protection from preening as well as warmth and humidity when seabirds are at rest. Also, this part of the wing is subject to the least air turbulence during flight, because seabirds have well-developed alulae, or wing slots. If you are a feather mite living on the surface of feathers, this is prime real estate. When mite populations grow very large, there can be stiff competition for this desirable area, with some species being more likely than others to be pushed out of their first-choice (business-class) habitat. For example, Choe and Kim found that when a particular species of *Alloptes* encountered *Laronyssus*, it got pushed into sites closer to the bird's body, where there are smaller feathers with shorter barbs. Over time, these competitive pressures can partition larger ecological niches into smaller ones, thereby promoting biodiversity of feather mites by forcing them to adapt to very specific niches.

Temperature, humidity, and the age of the feathers can also affect the distribution of feather mites,[21] and although the habitat distributions of feather mites are finely tuned, they can be subject to rapid change when conditions warrant it. For example, mites on the wings of blue tits (*Parus caeruleus*) are found on the birds' primary and secondary feathers in the summer, but when temperatures fall below 10°C in winter, they move closer to the body, onto tertiary feathers.

Feather age is also important but for different reasons. Old feathers have been on the bird for a long time, so mites have had time to colonize them—but old feathers are at risk of being worn, damaged, or moulted. In fact, one of the functions of moulting may be to unload parasites, a strategy that is widely shared with other animals and even plants; when reptiles shed their skins, or mammals shed hair, or plants drop their leaves, they may be trying to rid themselves of parasites.

Because feather mites live in and on feathers, moulting is a critical time in their lives. How do they survive this event? There are several possible scenarios: Perhaps mites do nothing and feathers are re-colonized after each moult. Perhaps mites randomly distribute themselves over feathers of varying ages, allowing some to escape moulting of older feathers. Or perhaps mites can somehow detect that a feather is about to be shed and take evasive action.

Roger Jovani and David Serrano analyzed distributions of feather mites on 63 moulting passerines and discovered that the numbers of mites were reduced on feathers that were next in line for moulting—in other words, mites evacuated the feather before it was shed.[22] Mites' ability to escape varied for the different types of feathers—they were able to escape from moulting primaries and secondaries better than from tertiary feathers. (This may be another factor that makes some areas more attractive as habitats, leading to competition for space.) In another study of surface feather mites (Pterodectes rutilus) on barn swallows (Hirundo rustica), researchers found that mites make precise decisions about the feathers they inhabit, and again, manage to escape from those about to be moulted.[23] It seems that feather mites are very adept at detecting changes in the vibrations of senile feathers and pull up stakes accordingly.

Overall, the distribution of mites on a bird at any time is a compromise between many factors: mites look for feathers that provide the best architectural match for their physique, where they are unlikely to be preened or blown away by air turbulence, where it is warm and humid, and where they are not likely to be abruptly evicted from their home as a result of moulting. On top of all this, mites must contend with other mites and with feather lice that may try to eat them or compete with them for homes.[24]

The lives of feather mites are intimately tied to their host birds, but what is the nature of this relationship? Are feather mites parasites that reduce the fitness of birds? Are they harmless commensals that have little impact on their hosts, or can feather mites even be mutualists that improve the lives of birds?

Rather than being armed with blood-sucking stylets, feather mites have mouthparts that allow them to feed on the debris and fungal spores stuck in preen gland secretions. Waterbirds and shorebirds, with larger preen glands for secreting waterproofing oils, have more vane mites than other birds such as passerines.[25] Even within a species of bird, there is a positive correlation between gland size and the abundance of feather mites,[26] suggesting that gland secretions are an important source of food for feather mites. In addition, mites on the skin surface probably feed on sloughed-off skin flakes, while quill mites eat the keratinous medulla of the quill or fluids from the papilla at the base of feathers.

This diet does not seem to cause hardship for host birds. Studies that show a negative, parasitic effect of feather mites on birds are rare. Chelsea Bueter did find that the abundance of feather mites on eleven species of birds was associated with reduced wing length, weight, and body fat—but it was not clear whether large mite populations were the cause or the result of these undesirable characteristics.[27] Feather mite populations can reach biomasses that approach 10% of the wing mass of their host,[28] but there is no clear evidence that this has a negative impact on the host.

More commonly, feather mites appear to have no measurable effect on the health or survivorship of hosts. For example, Seychelles warblers (*Acrocephalus sechellensis*) are infested by a species of *Trouessartia* mites in the Seychelles and Cousin Islands. Mite loads on warblers were not found to be related to body condition, time spent grooming, or survivorship, so the relationship can best be described as benign.[29]

It's conceivable, though, that housing a large population of feather mites could actually be good for birds. How could this be? In theory, mites provide a cleaning service, helping birds keep their feathers in tip-top condition by eating preen gland secretions, which contain pollen, algae, bacteria, skin debris, and fungal spores. They may even offer a pest control service—mites sometimes feed on the body fluid of lice and louse flies, which are known to be parasites of birds. In fact, several studies have found that having a large number of feather mites improves the condition and survivorship of birds.

Guillermo Blanco and his colleagues studied feather mites (*Gabucinia delibata*) on group-living, red-billed choughs (*Pyrrhocorax pyrrhocorax*) in Spain.[30] Fledglings of these communally breeding birds had no mites until they were one to five months old, when they left the nest and began to roost with other birds. Among choughs, an abundance of mites was an indicator of a healthy bird. The researchers also observed that cross-billed birds, whose preening was impeded, had the same number of mites as normal birds. This suggests that normal preening has little effect on the birds' feather mites—an outcome that is sensible for the birds if the mite–bird relationship is either commensal or mutualistic. If mites actually improve feather condition and protect birds from pathogens, it would be in their interest not to remove mites along with the noxious parasites that are controlled by preening.

Similarly, a study of feather mites (*Pteronyssoides obscurus*) on cliff swallows (*Petrochelidon pyrrhonota*) showed a pattern in which swallow survival increased with larger mite loads.[31] The researchers speculated that mites were removing preen gland oil that contained pollen and fungal spores, improving the condition of the birds' feathers. They may also have been competing with, or even eating, harmful bacteria and fungi. Studies like these, which link good health in birds with a large mite load, are very suggestive. But in order to conclusively show that some feather mites are mutualists and to get some insights into how they help birds, researchers need to do controlled experiments in which they intentionally vary mite loads and observe their effects on birds. Nevertheless, the possibility that birds benefit from their tiny tenants raises interesting questions. Just as birds have evolved behaviours that help them get rid of parasites, might they also have developed behaviours to increase their numbers of beneficial mites? For example, might birds intentionally interact to share feather mites?

Feather mites have physical adaptations that are harmonized with hosts. For example, mites on the outermost wing feathers are large and robust. They have well-developed muscles and strong and heavy exoskeletons. They reduce their height by being flattened and they have short dorsal setae (hairs), perhaps to reduce air turbulence. Their posterior legs project sideways (rather than underneath) and often have elongated setae that entangle the fine barbs of feathers. Sometimes, these mites have very asymmetrical appendages and setae that are fine-tuned to the feather structure where they live. In contrast, quill mites are adapted to live inside feathers by having longer, more cylindrical bodies, weaker musculature, and reduced sclerotization of exoskeletons. The fact that many different and unrelated species of mites share similar adaptations and are suited for the habitat where they live is great evidence of convergent evolution; if you are a mite and choose to live on the windswept surface of a flight feather or the interior of a quill, there are only a limited number of forms that your body can take.[32]

More complex, two-way relationships between birds and mites can also be observed, as in the case of the tropical fowl mite *Ornithonyssus bursa*. This blood-feeding parasitic mite stimulates immune responses when it bites, and, in order to counteract the host's attempts to disrupt its feeding, it secretes anticoagulants and anti-inflammatory chemicals in its saliva.[33] This has created a host–parasite arms race, just as we saw for ticks in Chapter 4. The fact that we see chemical adaptations, as well as adaptations of body shape, strongly suggests that mites are great candidates for co-evolution— including both co-adaptation and co-speciation. And, if evidence of co-evolution does exist, then we can use the evolutionary histories of mites to inform the histories of birds, and vice versa.

Some studies have compared the evolutionary histories of mites and birds. For example, Dabert and Ehrnsberger looked at the mites in the family Ptiloxenidae, and matched them to their grebe, stork, and shorebird hosts.[34] There was generally congruence between the evolutionary histories of mites and hosts, but some birds were missing mites. Perhaps some ancestors failed to acquire mites, or possibly mites on these birds have gone extinct. One unsolved puzzle relates to grebes and flamingos, which have been linked to a recent common ancestor, as discussed in Chapter 2 (and which will be revisited in Chapter 7).[35] Although both feather lice and internal parasites support this relationship,[36] so far, no evidence from studies of feather mites supports it—perhaps these birds and mites should be examined more intensively.

However, there are two fascinating studies that shed some light on a controversial problem in bird evolution.[37] These studies involve ratites, which are mainly flightless birds found in the southern hemisphere. The group includes extinct giant elephant birds and moas as well as their living relatives, including kiwis, emus, ostriches, rheas, tinamous (one of the few ratite species that can fly), and cassowaries. Cassowaries are interesting because they were not known to play host to any feather mites, even though other ratites are infested by many species—that is, until Heather Proctor picked up a feather of a southern cassowary, *Casuwarius casuwarius*, in Australia and discovered some unique mites. This mite stepped right into the middle of a debate over ratite origins.

Ratites have several characteristics that differ from most birds, including a flat, raft-like sternum (breast bone) and a unique palate (called a paleognathous palate). For many years, biologists thought the paleognathous palate was a very primitive feature. Because the mostly flightless ratites are distributed on islands and continents separated by large stretches of ocean, but share such a primitive feature, it was thought that they must have originated from a common ancestor that lived on the prehistoric island continent of Gondwanaland, about 180 million years ago. Consequently, as continental drift split off and moved parts of Gondwanaland over the southern hemisphere, isolated ratite populations were stranded on these continent-sized rafts. Later, they evolved to become the species we see today. According to this theory, Madagascar elephant birds (*Aepyornis maximus*) and African ostriches (*Struthio camelus*) are the oldest and most closely related ratites.

However, are ratites in fact primitive birds? Some studies have questioned this interpretation. For example, analysis of mitochondrial DNA showed that elephant birds were more closely related to kiwis than they were to ostriches.[38] Also, cassowaries and their kin evolved more recently, after the split between passerines (perching birds) and other *neognathous* ("typical" palate) birds.[39] This evidence supports a different idea, proposed in 1956, that the shared characteristics of ratites (that is, their large sizes and their juvenile features such as plumaceous feathers and flightlessness) stem from the fact that ratites live in similar habitats, rather than being the result of a common ancestor.[40] This implies that many features of ratites are due to convergent evolution and have more to do with their lifestyles than with their evolutionary relationships. Could mites provide any information that could clarify these ideas?

This is where Heather Proctor's mite—so far, the only one ever discovered on cassowaries—comes into the picture. It turns out to be a unique mite, called *Hexacaudalges casuaricolus*.[41] It has many special features and is the only representative of its family from ratites, suggesting that it originated on cassowaries or their ancestors long ago. The closest relative of *H. casuaricolus* is found on hornbills (Coraciiformes: Bucerotidae), which are medium-sized birds with many interesting features (including eyelashes, a huge, ornamented bill, and the weird habit of having moulting females sealed off in the nest by males). Hornbills occur in tropical Africa and southeast Asia. So, how did a mite that usually occurs on hornbills get onto cassowaries? Mironov and Proctor speculate that this mite is an antique and that most representatives of its ancient family have gone extinct. However, there is the possibility that in the distant past, hornbills and cassowaries shared a common habitat and mites shifted from one host to the other. More samples from birds in the southern hemisphere would shed light on this evolutionary puzzle, but for now, the mites of ratites seem to hint that they are not primitive birds. The mites of these birds confirm the story we are getting from DNA—ratites came from different ancestors and their flightlessness, gigantism, and other features evolved independently several times.

This chapter began with a famous proverb that was inaccurate regarding mites, so it only seems appropriate that it end with another proverb that better summarizes the role of mites in the flying zoo: "Don't ignore the small things—the kite flies because of its tail." Despite their small size (and probably because of it) mites are very important in the world of birds. From a scientific and natural history perspective, their importance far exceeds their size—they tell us tales about how birds live and what birds do. They can even tell us about the evolutionary history of birds, and if we keep our minds open enough to not ignore them, and if we are willing to study the smallest kinds of animals, mites can challenge us with an unending supply of intriguing questions.

6

F L Y I N G

Z O O   F L I E S

TINY PREDATORS

Have you ever visited a zoo and, as you stood entranced, looking at some beautiful exotic animal like a highland gorilla or a Siberian tiger, noticed that flitting around in the same enclosures were some English sparrows, and on the ground of the enclosures were burrows made by common ground squirrels? Perhaps you heard the chattering of a red squirrel up in a tree, arguing with a pigeon over the right to some perch. These ubiquitous, commonplace animals seem to ruin the natural biome effect that zookeepers try to achieve. They make it harder for you to imagine that you are in the middle of a Rwandan cloud forest or on a windswept plain in Central Asia. That's how I view most of the flies that are found in and around the flying zoo—they are transient, generalist, hit-and-run parasites—they don't really seem to belong as integral parts of a specific ecosystem within the flying zoo. It is little wonder that some biologists refer to them as micropredators rather than as parasites.

Whereas most of the residents of the flying zoo have spent a long time with their host birds and have travelled together along the same winding evolutionary pathway, many flies come and go, and they may be just as likely to get blood from a mammal or an amphibian or from you as they are from a bird. This picture, however, is incomplete and doesn't capture the full impact flies have on birds. When I imagine the thoughts that might preoccupy birds, I envision them spending an inordinate amount of time complaining and worrying about flies. Indeed, some flies have fully embraced the parasitic lifestyle, and, like lice and mites, have become important, fundamental parts of the flying zoo.

Flies belong to the insect order Diptera (*di*—"two"; *ptera*—"wings") and include animals such as mosquitoes, blackflies, tsetse flies, and louse flies. Although dipteran flies do have two pairs of wings, one pair has become modified into small, club-shaped structures called halteres, which act as gyroscopes that provide flies with information on roll, pitch and yaw. Flies also tend to have narrow "necks" and "waists"; overall, they have the typical head, thorax, and abdomen sections we associate with most insects. Their mouthparts are modified to take advantage of feeding on coarse tissues such as feathers or on liquid tissues such as blood, a characteristic that is probably especially important for parasitic flies.

Flies have a life cycle with complete metamorphosis: that is, eggs hatch to produce grub-like larvae (often called maggots), which become pupae, and, eventually, winged adults. In a few species, such as louse flies and tsetse flies, eggs remain inside the adult female until hatching; these species can be considered viviparous (giving birth to live young). Some species, such as botflies, are more insidious and parasite-like, and they deposit their eggs into a bird so that the larvae grow and develop inside. The thought of maggots squirming and growing inside a live bird may make many of us—as well as their hosts—very uncomfortable.

The dipterans known as biting flies are more like micropredators. These include the tiny flies commonly known as punkies, midges, or noseeums that belong to the family Ceratopogonidae, blackflies (family Simuliidae), and mosquitoes (family Culicidae). In addition to the blood loss and irritation they cause, these small predators transmit viral, bacterial, protozoan (including malaria), and animal parasites.

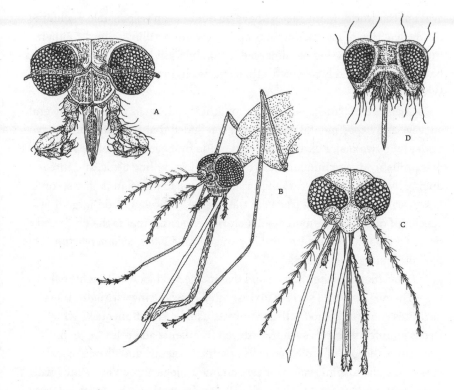

FIGURE 6.1 The mouthparts of flies are modified to allow for different styles of feeding.
(A) A blackfly (*Simulium*). (B) A female mosquito feeding. (C) Frontal view of a female mosquito's mouthparts, beginning from left to right next to the antenna: mandible, maxillary palp, maxilla, hypopharynx, labium with a spiny labellum at its tip, labrum, labial palp, and the other antenna.
(D) A louse fly (Hippoboscid); a diagram of the entire fly appears in Figure 2.3.

MALARIA AND ITS CARRIERS

Captain James Cook was the European discoverer of the Hawaiian Islands in 1778. At that time, the islands were indeed a paradise, with lush forests and many unique birds endemic to the islands. Since then, a larger proportion of birds has gone extinct in the Hawaiian Islands than in other comparable areas of the world. Many factors have no doubt been responsible, including the destruction of natural habitats and the wanton killing of birds. But an interesting pattern that has emerged is that birds living at higher elevations have been more likely to survive than species living at lower elevations. Why is this?

Charles and Sandra van Ripper and their colleagues conducted several surveys and experimental studies, and concluded that Hawaiian birds have undergone two major extinction events.[1] The first occurred before 1900 and was probably due to humans destroying habitats, killing birds, and introducing invasive predators, such as rats and cats, to the islands. The second, however, can probably be traced to the release of foreign birds, including a species of quail (the painted quail, *Coturnix chinensis*) that is the preferred host of a blood-dwelling, protozoan parasite called *Plasmodium relictum*. This parasite causes bird malaria.

P. relictum parasites invade and live inside red blood cells, where they ingest hemoglobin and secrete toxins. After reproducing asexually, these parasites cause red blood cells to rupture, usually simultaneously, which floods an infected bird's system with cell fragments and releases toxins. Over time, the loss of red blood cells results in anemia and physiological stress, and the toxemia causes fevers and neurological problems. The bird's liver must work so hard to detoxify cellular debris that it becomes enlarged, and the spleen also becomes enlarged and distended as it attempts to compensate for the loss of blood cells. In addition to these acute effects, bird malaria can have longer-term, hidden costs.[2] In reed warblers (*Acrocephalus arundinaceus*) in Sweden, mild chronic malaria was associated with degradation of the ends of chromosomes, called telomeres. This accelerated shortening of chromosomes was linked to reduced lifespan in these birds and reduced numbers and quality of their offspring. In hosts with no previous experience with *Plasmodium*, such as the defenseless birds endemic to Hawaii, the consequences can be devastating.

Malaria parasites are transmitted by mosquitoes. In Hawaii, the culprit is a North American species called *Culex quinquefasciatus*, which has a predilection for biting birds. It was likely introduced to the islands in 1826. Surveys of birds caught on Mauna Loa showed that more than 40%

of some bird species were infected with malaria, and that native birds had the highest percentage of infected individuals. Native birds also had the highest numbers of infected red blood cells, indicating that they were more susceptible to infection. Birds from wet mesic forests were more highly infected than birds from dry xeric forests, probably because the former provide ample breeding sites for mosquitoes. The greatest parasite levels were observed in September and October. Mosquitoes generally had lower densities at higher elevations but could still breed there during the warmest months of the year. In short, native birds were at risk of contracting malaria at any time and in any location, but the greatest danger of transmission occurred in mid-elevation wet forests. Native birds were never found at lower elevations and most lived at elevations higher than a thousand metres. The more time birds spent in mid-elevation wet forests, the more likely they were to become infected with malaria. It seems likely that malaria has forced present-day native Hawaiian birds to live in high-elevation forests, restricting their habitats.

Birds living on isolated islands are rarely exposed to most of the predators, parasites, and infectious diseases that are regular aspects of the daily life of mainland birds. Consequently, island birds are frequently immunologically naïve. The introduction of a parasitic infection like malaria by way of an introduced host species is a double whammy—native birds easily succumb to the disease, which then creates an ecological opening for the invading foreigners to exploit, allowing these to expand their ranges and provide stiff competition for native birds.

In Hawaii, a beautiful red and black honeycreeper called the iiwi (*Vestiaria coccinea*) is the native bird that is most susceptible to malaria. Not surprisingly, it has gone from being one of the most common forest birds to being rare or locally extinct. In contrast, the common amakihi (*Loxops virens*), which is the most malaria-resistant native bird, occurs in greater numbers and has even managed to recolonize areas from which it had disappeared. The researchers studying them have suggested that amakihi populations do not migrate to higher altitudes and back, and would therefore be exposed to malaria more regularly, which would result in selection for the most resistant birds.[3]

Even the behaviours of native Hawaiian birds have probably been altered by mosquitoes and the malaria they carry. In a 1968 study of captured native birds, Richard Warner claimed that the birds slept with their bills, faces, and legs exposed, whereas introduced birds tucked their heads into their fluffed back feathers and covered their legs with their bodies.[4] But by the time the van Rippers did their study in 1986, the captive amakihi, iiwi, and another honeycreeper, the apapane (*Himatione sanguinea*), all slept with their bare skin protected from biting mosquitoes.

Honeycreepers and other Hawaiian birds that feed on nectar must follow the altitudinal patterns of the flowering of trees. Nectar-producing trees flower at lower elevations in the summer and fall, and then gradually flower upslope, with trees at the highest elevations flowering in winter. Many birds are therefore forced into lower-elevation forests in the fall, just when the populations of mosquitoes are expanding. This means that in order to forage optimally, these birds must risk exposure to malaria.

In an attempt to reduce their exposure, apapane and iiwi now undergo daily migrations. During the day, when the mosquitoes are less likely to bite, the birds move downslope. They migrate back upslope just before dusk. The daily migration of honeycreepers closely matches the biting activity of mosquitoes—by the time that *Culex quinquefasciatus* is ready for a blood meal at about eight in the evening, all the smart birds have left. Over a period of about fifty to a hundred years, what was once a gradual, seasonal migration has now evolved into a daily event under the selection pressure of mosquito-borne malaria.

Another example of a bird behaviour triggered by malaria-carrying parasites can be found among great horned owls (*Bubo virginianus*), which are large, long-lived forest owls that live in North and South America. In the boreal forest of the Yukon, in Canada, Christoph Rohner and his colleagues were puzzled when they noticed that great horned owls in the post-fledging stage roosted in sunny, exposed sites on the ground, while adult owls, like other forest owls, always roosted in concealed positions in trees, at four metres above the ground or higher.[5] Solitary adult owls behaved like family groups (that is, roosting in trees), so there did not seem to be any special behavioural reason for juveniles to roost on the ground, nor were ground-roosting owls feeding there.

Within their study area, the researchers discovered that bird-feeding blackflies (Simuliidae) transmit a malaria-causing protozoan called *Leucocytozoon ziemanni*, which results in anemia and has been identified as a major cause of juvenile owl mortality. Could the unusual roosting behaviour of young owls, they wondered, be a response to blackflies?

Birds often choose roosting sites that will conceal them from potential prey or predators, prevent them from being mobbed by other birds, or offer some thermal advantage that helps the birds save energy. Rohner and his colleagues found that great horned owls shifted their roosting sites from the mid-canopy level in trees to exposed sites on the ground in order to drastically reduce their exposure to blackflies and their risk of contracting malaria.

Bird malaria caused by *Leucocytozoon* is not a rare or harmless disease. In areas ranging from California and New Mexico to Finland, owls (including spotted owls, *Strix occidentalis*, and Tengmalm's owls, *Aegolius funereus*, also known as boreal owls) can have infection rates of 95%, with malarial infection resulting in smaller clutches, anemia, and high rates of mortality. In addition, the direct effects of blood feeding by the black-flies can cause external lacerations, anemia, and lower survival rates.[6] It is no wonder that great horned owls would trade the nicely concealed mid-canopy sites, where they are tormented by disease-causing blackflies, for the respite of a sunny, exposed ground site. In fact, we might wonder why great horned owls in the boreal forest don't spend even more time at ground level.

The researchers suggested that the owls' decision to roost on the ground revealed an intricate cost–benefit equation: Great horned owls could also avoid blackflies by roosting at the tops of trees, but here they were harassed by gulls and shorebirds, and attacked by hawks and ravens. On the other hand, roosting on the ground exposed the owls to dangers from larger predators.

At the Yukon study site, the main food of owls (and many other predators) was snowshoe hares, which have populations that fluctuate cyclically over ten-year periods. Rohner and his colleagues reported that, in years when hares were scarce, lynx and coyotes captured and ate many juvenile great horned owls roosting on the ground. This observation suggests that there is a trade-off for owls between the risk of being eaten by mammal predators and the risk of being attacked by micropredators such as black-flies. Once again, birds and their behaviours appear to be shaped by a complex set of interactions in their ecosystem.

BLOWFLIES AND TASTY CHICKS

Many of the flies in the flying zoo move around as adults and take blood meals from several birds. But some also feed as larvae and are thus restricted to feeding on one host—often a chick and its nestmates. Blowflies (Calliphoridae) lay their eggs in the nests of a wide variety of birds. When the eggs hatch, the larvae become nocturnal parasites that take blood meals throughout the night but burrow down in the nest during the day. Adult flies often have a blue or green metallic sheen, and are known as bluebottles, greenbottles, or even "blue-arsed" flies (in Britain and Australia). The term "blowfly" probably comes from the fact that these flies are adept at finding carrion, and maggot-riddled meat was known as "fly-blown."

Parasitic *Protocalliphora* blowflies are blood-feeding pests that occur in the nests of a variety of birds. Adults lay eggs in nests or directly on birds. The eggs hatch quickly, within 24 to 48 hours after being deposited. Larvae feed repeatedly as they go through their developmental stages, taking anywhere from 7 to 15 days to reach the non-feeding pupal stage. The pupae spend between 9 and 36 days in the nest, depending on temperature and humidity, before adult flies emerge.[7] It is likely that only adults overwinter. This life cycle means that blowflies must time their reproduction so that larval stages coincide with the time at which nestlings are present, but before the young birds are fully feathered and cannot be bitten.

It is not unusual for nests to be heavily infested with blowflies. In a study of tree swallows (*Tachycineta bicolor*) in northern British Columbia, almost 90% of 37 nest boxes contained *Protocalliphora* larvae.[8] Infested nests contained an average of 50 larvae, and one nest had 178—generally, nests with larger broods have more blowflies. Nest infestation rates of 100 larvae per nest are not uncommon among a variety of passerines.

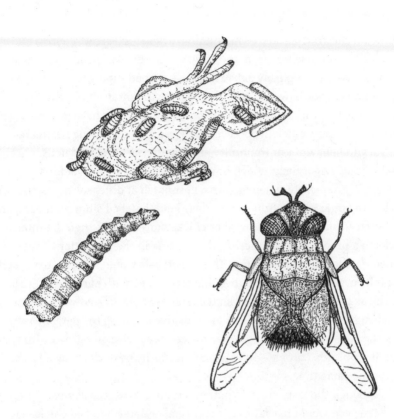

FIGURE 6.2   Blowflies at various stages. Adults lay eggs in bird nests. After hatching, the fly larvae hide in the nest material during the day but feed on blood and tissues at night. Here, a purple martin (*Progne subis*) nestling is attacked by several blowfly larvae.

Individual blowflies take a great toll on their hosts, perhaps greater than any other ectoparasite. For example, Terry Whitworth and Gordon Bennett estimated that as few as 5 larvae of *Protocalliphora chrysorrhoea* could drain all the blood from a barn swallow chick (*Hirundo rustica*) in just one day.[9] They also estimated that 360 larvae of *P. asiovora* could exsanguinate a larger magpie chick (*Pica pica*). Their estimates may be on the high side, but there is no doubt that larvae can wreak havoc on chicks; another study using a more conservative estimate indicated that as few as 8 to 20 *P. hirundo* larvae, feeding on chestnut-backed or mountain chickadees (*Poecile rufescens* and *P. gambeli*) took anywhere from 10% to 25% of the chicks' blood—a parasite load of 100 could completely exhaust a chick's blood supply.[10]

Not all chicks, however, receive equal treatment from blowflies or other ectoparasites. Phillippe Christe and his colleagues have proposed what they call the "tasty chick hypothesis," suggesting that, among nestmates that vary in behaviour or immunity, those with the weakest defenses are attacked most vigorously, which improves the odds of survival for stronger nestlings.[11] Frequently, when eggs in a nest hatch at different times, the first chicks are larger and healthier than their later-hatched siblings. If ectoparasites such as blowflies attack weaker chicks more readily, then the inferior chicks in a brood might be sacrificed, resulting in the protection of their earlier-hatched, hardier siblings. This hypothesis would receive strong support if it turned out that surviving chicks in a nest with parasites were as healthy as chicks from nests with no parasites at all—if that were the case, it would suggest that weaker chicks are effective at luring the parasites away from their stronger siblings, whose chances of survival are greater from the start. It would also provide an explanation of why birds don't lay all their eggs at the same time.

Although the tasty chick hypothesis is attractive, it is devilishly hard to test. First, you would have to show that smaller, later-hatched chicks have weaker anti-parasite defenses. Second, you would have to show that parasites treat chicks unequally and prefer to attack smaller, weaker chicks— that is, that the smaller, weaker chicks are somehow more "tasty" to the parasites. Third, you would need to demonstrate that first-hatched, larger chicks gain a significant benefit in survival from reduced parasite attacks, and that the ability to fend off parasites is a genetically controlled characteristic. Finally, you would need to determine that parasites occur frequently enough with birds to act as a selective pressure. Several alternative explanations are possible—for example, perhaps larger chicks physically force smaller siblings to spend more time at the bottom of the nest, where parasites have better feeding access.

Nevertheless, since it was first proposed, the controversial tasty chick hypothesis has stimulated lots of research, with researchers trying to determine whether some chicks make more attractive targets for micro-predators than others. A study of blowfly attacks on nestling blue tits (*Parus caeruleus*) used the turnover of radioactive insulin as an estimate of blood loss; this measure showed that a chick's mass and its relative size within the brood was correlated with the severity of blowfly attacks—scrawny chicks that ranked below their nestmates in size lost the greatest amount of blood.[12] Therefore, it appeared that blowflies did concentrate attacks on the weakest (tastiest) chicks. However, another study of blue tits found no evidence that blowflies concentrated their attacks on last-hatched chicks—in this case, there were no "tasty" chicks.[13]

Alexandre Roulin and his colleagues found that support for the hypothesis varied depending on the fly in question (blowflies versus louse flies) and the method of measuring the strength of chicks' immunity (humoral responses, which involve antibodies, versus cell-mediated responses, which recruit T cells to attack pathogens).[14] The blowfly, *Carnus haemapterus*, selectively targeted later-hatched chicks in barn owl (*Tyto alba*) and kestrel (*Falco tinnunculus*) nests. Moreover, first-hatched barn owl chicks had the strongest humoral immune responses. These pieces of evidence support the tasty chick hypothesis. However, other evidence contradicted the hypothesis: louse flies (*Craterina melbae*) infested mostly *earlier*-hatched chicks among alpine swifts (*Apus melba*) and great tits (*Parus major*), and there was no difference in cell-mediated immunity among chicks, regardless of their hatching order. Roulin and his colleagues concluded that overall, the tasty chick hypothesis did not receive broad support, a position that was echoed by Roberto Valera and his colleagues.[15]

It turns out to be difficult to make any broad, sweeping conclusions about flies and their effects on their hosts. At times, it may be hard for researchers to detect any effects at all of blowflies on their hosts. Although many studies have shown that blowflies can drain enormous amounts of blood from their young hosts, leading to reduced hemoglobin levels, anemia, and poor growth and survival, not all studies have been able to find a connection between blowflies and the health of their hosts. This puzzle may be explained by the fact that many birds seem to physiologically adjust to the costs imposed by these parasites; they may increase their production of red blood cells, accelerate their growth at the end of the nestling period, or delay fledging. In addition, infested birds may pass on the costs to their parents, by begging for food more frequently. Still, this may not be enough to offset the damage; in one study carried out in Corsica, researchers found that even though parents in infested nests worked harder to feed their chicks, the chicks that survived were smaller, lighter, and sicker than chicks from clean nests.[16]

If there is one important ecological lesson that comes from studying flies, it is that their effects on birds result from the interplay of several factors, including the flies' parasitic properties, the health of the host bird, and even external conditions such as food availability and environmental temperature.

LESSER OF EVILS

The lesson of ecological complexity is revealed even more dramatically by another kind of dipteran that attacks birds—the botfly. Botflies (genus *Philornis*) belong to the same family as the all-too-familiar houseflies (Muscidae). Like blowflies, bots are not parasitic as adults, but their larvae, or maggots, live inside the host and feed.

In birds, maggots tend to be found between the muscles and the basal layers of skin. Maggots are armed with spines that serve two purposes: the spines make it difficult for hosts to pull them out of their connective tissue cysts, and, as the maggots wriggle inside, the spines tear tissues, causing inflammation and the leakage of serum and blood, which serve as food. A friend of mine, whose shoulder was infected by mammal botflies he acquired in the lowland rainforest of Ecuador, told me that most of the time, he was not aware he was infected, but every now and then, when the bots wriggled, he felt a sharp, stabbing pain.

Within a cyst, the bots breathe through a respiratory pore in the bird's skin. Cysts can occur just about anywhere, but most often, they are seen on the bird's head or in the remigial areas of the wings. Botfly larvae have been found in the ear canals and beaks of eastern bluebirds (*Sialia sialis*).[17] After about four to six days, mature maggots (now about 1.5 cm long) leave the cyst

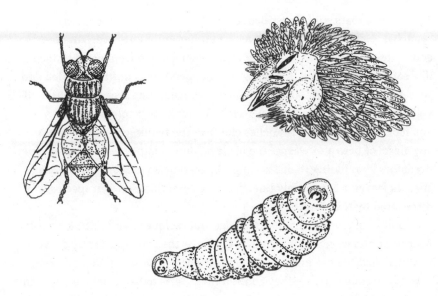

FIGURE 6.3   Botflies (*Philornis downsi*) at various stages. During the larval stage, mouthparts and body spines cause the bird to release blood and serum, which nourish the maggot. A nestling of a crested becard (*Pachyramphus validus*) with a botfly larva in a nodule just below its eye. Note the breathing hole needed by the larva.

through the respiratory pore, fall off the bird, burrow into the substrate of the nest, and secrete a cocoon within which they pupate.

Like blowflies, botflies are serious parasites of birds. They are associated with slowed growth, reduced body mass, and delayed fledging—and when infections are intense, they kill. In fact, the havoc caused by botflies may be so great that it has forced some birds to accept a distasteful protective measure—voluntary invasion by other parasites.

One such case can be found along the Tiputini River, which winds and twists its way through the lowland Amazon rainforest in eastern Ecuador, meandering through some of the most pristine and biologically rich habitat left on Earth. The area has more than 1,500 species of trees (averaging 300 per hectare), more than 520 species of birds, 12 different primates, close to

80 species of bats, and an estimated 60,000 species of insects! Large tree branches, covered in mosses, ferns, and other epiphytes, overhang the chocolate-coloured water of the river. Hanging down from some branches, like softballs suspended in netting, are the pendulous nests of crested and russet-backed oropendolas (*Psarocolius decumanus* and *P. angustifrons*) and yellow-rumped caciques (*Cacicus cela*). The birds nest colonially, and as you look along the same branches closer to the trunk, you often see massive nests of bees and wasps. The birds no doubt gain some protection from predators by building their hanging-pouched nests in these precarious sites; however, a completely unexpected interaction between species was discovered by Neal Smith.[18]

Smith discovered that oropendolas and caciques that built their nests near bee nests were less infested by botflies—the bees apparently drove off the bots. Birds whose nests were not near bees, however, used a different strategy to control bots—they allowed their nests to be parasitized by giant cowbirds (*Molothrus oryzivorus*). Giant cowbirds are brood parasites. When cowbird young hatch in the nests of other birds, they can out-compete the resident nestlings for food, and in some cases, will even commit infanticide by tossing their hosts' nestlings out of the nest. Why would oropendolas allow cowbirds to remain in their nests?

Cowbirds preen themselves and their nestmates, thereby cleaning the nest of botflies. Moreover, it appears that botflies prefer to lay their eggs on cowbird nestlings rather than on caciques or oropendolas. According to Smith, when caciques and oropendolas nest in trees without bees, they accept cowbird eggs because they have no other protection against botflies. And indeed, oropendolas have greater success in producing offspring when parasitic cowbirds are present.

There is an environmental twist to this intricate ecological story. Recently, the Tiputini area has been opened for oil exploration. With oil exploration and discovery come roads, seismic lines, and pipelines. This attracts more people, who clear the rainforest. Cowbirds, which are very adaptable, seem to be able to invade these disturbed areas more easily than areas of pristine rainforest. How will this affect the oropendala–botfly–cow bird interaction? Only time and intensive ecological study will tell.

FIGURE 6.4   Three russet-backed oropendola (*Psarocolius decumanus*) nests hang from a branch near a bee nest. Giant cowbirds (*Molothrus oryzivorus*) lay eggs in the nests and parasitize the oropendolas. Botflies (*Philornis downsi*) lay their eggs on chicks in the nest and the larvae form capsules under the skin.

LOUSE FLIES AND THEIR CARGO

The last—and perhaps most curious—flying zoo flies are called louse flies, or hippoboscids. Louse flies are the most specialized parasitic flies that attack birds. They are not related to lice—they are true dipterans. When viewed from above, they resemble houseflies, but with spindly legs (see Figure 2.3). At the ends of their legs, they have large, recurved tarsal claws that help them cling to their host's feathers. Viewed from the side, it is obvious that hippoboscids have flattened tops and bottoms, and with a bit of imagination, they resemble the shape of US Air Force F-117 stealth fighter jets.[19] Hippoboscids that attack mammals (and are known as keds) often eject their wings, but those on birds keep their wings—at least for a while.

After stealthily arriving on a host bird, a louse fly uses its flat body to move through the feathers and get down to the skin surface, where it deploys its needle-like mouthparts to suck blood. Hippoboscids demonstrate some degree of host specificity, but they are mobile and can attack other birds of the same species or, sometimes, of other species. They vary in size, so smaller louse flies use smaller hosts, and larger louse flies occupy larger birds—on hawks, for example, louse flies may be up to two centimetres long. In one of the most graphic and memorable descriptions of this relationship ever written, Miriam Rothschild and Theresa Clay wrote that "a small bird with one or two of these insects creeping about in its feathers can be compared to a man with a couple of large shore crabs scuttling about in his underclothes."[20]

The reproductive biology of louse flies is unusual in that females never lay eggs. Instead, a fertilized egg hatches within the uterus of a female fly and is fed by special glands. The larva completes its moults within a female and is deposited on the ground (or in a nest), where it pupates. Louse flies have one generation per year and produce very few young. To compensate for these limitations, adult female flies live long lives, nestled in the plumage of their host. The low reproductive rate of hippoboscids (an unusual characteristic for parasites) may be an adaptation to prevent over-infestations of their bird hosts.

Besides their nasty habit of drinking blood, hippoboscids affect birds by transmitting microparasites like malaria, trypanosomes (flagellated protozoans that live in the blood), and probably viruses such as West Nile virus (which will be discussed further in Chapter 10). Moreover, hippoboscids have been cited for years as agents of dispersal for chewing lice and mites—an ecological interaction called phoresy, discussed in Chapter 2. (However, not all mites or lice may avail themselves of this form of transport; one study of alpine swifts and feral pigeons failed to find feather mites attached to any of more than 400 species of hippoboscids, although the researchers did find skin mites catching rides.[21])

As you may have guessed, some mites have taken their relationship with louse flies one step further, and are now parasites of the parasites—that is, they are hyper-parasites. Some epidermoptid mites have a life cycle in which males are parasites of birds, but adult female mites embed themselves near the end of the abdomen of a louse fly and suck the fly's blood. In this case, bird-parasitic mites may be doing a good deed for birds by bothering the louse flies that attack them—although this act of charity is mitigated by the fact that female mites lay their eggs on louse flies, which hatch and get onto the bird. These mites then burrow under the superficial layers of the bird's skin, causing dermatitis, commonly known as mange.

Interestingly, louse flies can become drivers of mite evolution. For example, in a study of birds in the Galapagos Islands, flightless cormorants (*Phalacrocorax harrisi*) and Galapagos hawks (*Buteo galapagoensis*) were infested by what appeared to be a single species of skin mite (*Myialges caulotoon*) transmitted by two types of louse flies (*Olfersia sordida* and *Icosta nigra*). However, analysis of mitochondrial DNA of these mites revealed that there were really two kinds, each of which was restricted to the particular hippoboscid fly upon which they hitched rides, but not to the species of bird they infested.[22] Because the life histories of louse flies correspond closely to those of their hosts, this ecological relationship may be leading to the formation of two new species of mites in the Galapagos.

DEFENSIVE ACTION

At first glance, flies might seem like unspecialized intruders that flit in and out of the flying zoo without being an integral part of the ecosystem. But they are serious and significant members that have had an effect on the environments they inhabit. Birds have had to develop adaptations to prevent, or at least reduce, exposure to flies. In Hawaii, flies have altered the behaviour and feeding migration patterns of birds. In Canada, they have forced great horned owls to reconsider their choices of safe roosting locations. In the tropics, flies have compelled some birds to allow danger-ous brood parasites in their nests. Birds try to mitigate the effects of flies by feeding their young more food, laying their eggs at different times, altering fledging times, or re-nesting.

Sometimes, birds' defenses against flies come with wide-ranging consequences. For example, European starlings (*Sturnus vulgaris*) appear to have developed defenses against mosquitoes (*Culesita melanura*) that are not entirely benign. Mosquitoes develop preferences for feeding on specific hosts, and starlings are not their top choice. Researchers found that when European starlings and robins (*Turdus migratorius*) were equally abundant, mosquitoes contained lots of robin blood and little starling blood.[23] They speculated that mosquitoes prefer feeding on robins because starlings use a vigorous defensive behaviour that interrupts feeding (mosquitoes can only "nibble" on starlings), while robins don't interrupt the blood meal (mosqui-toes can "feast" on robins). This helps starlings to avoid being blood donors to mosquito recipients, but it can have major consequences for disease transmission.

Both hosts can be infected with eastern equine encephalitis virus, but starlings are more serious carriers of the virus than robins; when robins are maximally viremic, they will only infect about half the mosquitoes that take their blood, while starlings will infect almost all mosquitoes. The starlings' tendency to interrupt mosquito feeding before they are full of blood only increases infection rates, because this behaviour encourages mosquitoes to bite more birds. A similar system may be occurring with West Nile virus in North America, where bird defenses are increasing the spread of the virus.

Birds have a variety of anti-fly defenses at their disposal. They may smear chemical repellants such as crushed ants or plants onto their plumage or in their nests—these serve as general anti-ectoparasite agents. Pitohuis in New Guinea emit a pungent, sour odor and have a chemical in their feathers that can kill lice and that may also repel flies.

In the North Pacific, crested auklets (*Aethia cristatella*) make use of potent bug repellant. These auklets are comical-looking seabirds with red beaks and a showy plume of black feathers at the front of their faces. The birds nest in burrows and feed at sea on zooplankton, which they store in gular pouches below the lower mandible of the bill. Some auklet colonies in Alaska are close to wet tundra, and wet tundra in the north is ideal habitat for the production of huge numbers of biting flies, including mosquitoes. Hence, defensive action is needed.

If you are ever lucky enough to visit a crested auklet colony during the breeding season, prepare to be assaulted by the smells of regurgitated squid, bird feces, and an acrid odor secreted by the birds that includes octanal and hexanal aldehydes. In their colony, crested auklets emit almost pure aldehydes. Hector Douglas and his colleagues claimed that these chemicals prevent mosquitoes from making contact with birds; in experiments with captive mosquitoes, the aldehydes were as effective as commercial repellents that include DEET (N,N diethyl-meta-toluamide).[24] These auklet secretions may reduce blood feeding by mosquitoes and other ectoparasites such as ticks and may also reduce the resulting transmission of blood-borne infections.

Colonial nesting birds generally pay a high price to ectoparasites due to their lifestyle; the large number of birds in one place attracts biting flies, the re-use of nests promotes large populations of nest-dwelling parasites to build up, and the high density of birds permits easy transmission of parasites between them. Nest-based ectoparasites such as blowflies tend to have aggregated distributions in which some nests are heavily infested but most nests have few parasites. This has probably led to an interesting anti-parasite defense—adoption.

Adoption of unrelated young occurs when young birds leave their family nest during the rearing period to seek adoption by a neighbouring family.[25] Although this event has often been noted among colonial birds, the reason for it has been a puzzle. Pierre Bize and his colleagues have suggested that adoption is driven by ectoparasites.[26] They predicted that offspring from heavily infested nests would seek adoption, and that successful adoption would reduce their exposure to parasites. They tested their ideas by experimentally increasing and decreasing infestations of a blood-sucking louse fly, *Crataerina melbae*, in nests of colonial breeding alpine swifts (*Apus melbae*) in Switzerland. Swift nestlings in heavily parasitized nests (17 louse flies per nestling) switched nests 6 days earlier than lesser-infested birds (7 louse flies per nestling), and they switched nests 1.6 times more frequently. By abandoning their home nests for more attractive ones, the nestlings reduced their ectoparasite loads by 42%—adoption was indeed a successful strategy for reducing exposure to louse flies. The results of this study imply that ectoparasites of colonial nesting birds may be responsible for the emergence of certain social interactions, but it also leaves us with some questions: Why should foster parents accept the adopted young? Why do these birds nest colonially? Does this strategy only apply to very mobile ectoparasites (like louse flies) or is it more generally useful?

Flies in the flying zoo are important residents that affect many aspects of bird life today. In fact, flies likely affected the ancestors of present-day birds—possibly even before there were birds—and their competitors. Biting flies have been around for at least 142 million years,[27] and they probably carried disease-causing agents. For example, George Poinar reported that a hundred-million-year-old biting midge, preserved in amber, contained malaria.[28] Dinosaurs went extinct about 65 million years ago. This event is often attributed to a catastrophic event such as an asteroid collision or global climate change due to widespread volcanic activity. However, the extinction of dinosaurs was not quick—they suffered a lingering death. Could it be that malaria-transmitting flies were a major contributing factor to the demise of dinosaurs? The ancestor of modern birds likely lived sometime in the late Cretaceous. In the 65 million years since the extinction of the remaining dinosaurs, this ancestor has flourished, branching out into the amazing diversity of birds that we treasure today—possibly with the help of parasitic flies!

7

T H E   W O R M S
T H A T   A T E
T H E   B I R D

A DREAM HOME AND ITS RESIDENTS

Imagine you have found a new home. This home is very roomy, and its temperature and humidity are set to levels that assure you are perfectly comfortable—and these conditions never fluctuate, but remain stable, day in and day out, for your entire life. Imagine also that at your every whim, you are served an unending supply of water and the very best food you can eat, right in the perfect comfort of your home. Imagine that most of the worries of life—being killed in an accident, being attacked by a predator, getting caught in an unexpected storm—are of no concern to you as you stay safe inside your wonderful mansion. Now imagine all this is provided for free! This idyllic life is no mere fantasy. It is the lifestyle of an amazing array of animals that live inside birds: parasitic worms.

Most parasitic worms belong to one of three different phyla of invertebrate animals: the phylum Platyhelminthes (*platy*—"flat";

helminthes—"worms"; commonly known as tapeworms and flukes), the phylum Nematoda (*nema*—"thread"; called roundworms), and the phylum Acanthocephala (*acanth*—"thorn"; *cephala*—"head"; known as spiny-headed worms). These animal parasites are often lumped together and simply called helminths. Given the many advantages of taking up residence inside a bird, it is not surprising that each of these phyla has diversified, resulting in thousands of different species, or that within any one individual bird we may find more than a hundred thousand helminth worms. Nature does indeed abhor a vacuum, and the offer of free room and board for life is too good to pass up.

Birds are more mobile and they have broader diets and more complex organ systems (particularly their alimentary tracts) than most vertebrates, so, as a general trend, birds have richer, denser, and more diverse helminth communities than any other hosts.[1] Some species of worms are more successful than others; helminth species with the highest prevalence (infecting the largest number of hosts within a bird population) also tend to have the greatest abundance (occurring in the greatest numbers within any individual host).

This trend scales up. Frequently, the most abundant worms are also found in a number of different species of birds (that is, they have lower host specificity).[2] It seems that the characteristics that allow a worm to successfully infect a bird may also promote colonization and exploitation of other species of birds (at least if the hosts are moderately related and have similar food habits, as is the case with, say, different species of waterfowl). This trend also holds for most free-living organisms—species with good colonizing ability that can thrive in a variety of habitats tend to be widely distributed geographically. But interestingly, the patterns seen for bird parasites do not occur for fish parasites—in fish there appears to be a trade-off, so that if a parasite can infect numerous species of fish, it generally cannot achieve high population sizes.[3]

Al Bush and Clive Kennedy have pointed out that worm species that are flexible about their hosts and that are able to establish large populations gain insurance against the risk of extinction.[4] These characteristics also provide a lot of opportunities for the species to become genetically shaped

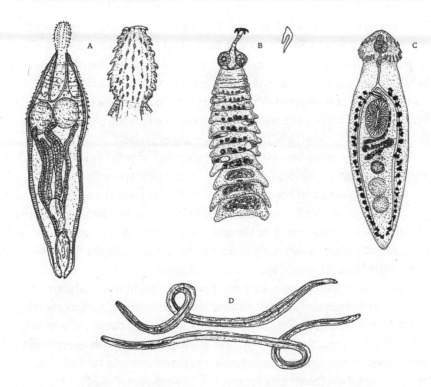

FIGURE 7.1   Representative examples of the main types of parasitic worms found in birds.
(A) *Polymorphus* is an acanthocephalan, or spiny-headed worm. It uses its retractable proboscis,
armed with rows of hooks, to attach to the intestines of its host, usually an aquatic bird.
(B) *Tatria* is a tapeworm, a type of flatworm. It attaches to its host, usually a grebe, with a
retractable proboscis armed with hooks, as well as four muscular suckers. (c) *Echinostoma*,
common in waterbirds, is another type of flatworm known as a fluke, or trematode. It attaches
using spines on a collar, a muscular sucker around its mouth (an oral sucker), and a larger,
ventral sucker or acetabulum. (D) *Contracaecum* is a roundworm, or nematode. Nematodes can
burrow into the lining of the intestine or crawl upstream to stay in place in their hosts, usually
fish-eating birds. Above is a male and below it, a larger female.

through interactions between members of the same species (for example,
through intraspecific competition), and for interactions to occur among different species of parasites (in interspecific interactions such as competition
or mutualism).

Intraspecific competition is well known for helminths—it manifests itself as a "crowding effect," in which individual worms that belong to a large population inside a bird are relatively small in size.[5] Interactions between species, such as selective site segregation, where different helminths sort themselves into different locations inside a host to reduce overlap among them, have also been demonstrated among parasites in the intestines of waterfowl[6] and in several species of grebes.[7]

Just as tiny mites are able to exploit a myriad of possible habitats within the plumage of a bird, helminths—with their generally small size—have a large selection of potential homes inside the bird. The chambers and passageways of the bird's lungs and air sacs can serve as a home, as can the urinary and genital systems, or the arteries and veins of the cardiovascular system. But without a doubt, most helminths reside inside the digestive systems of birds, as this site is the easiest to colonize.

Originally, free-living ancestors of parasitic helminths likely lived in habitats that contained animals commonly eaten by birds, such as aquatic invertebrates. Bird parasite ancestors probably started out as parasites of these invertebrates. Consequently, they would routinely be eaten by birds. Over evolutionary time, the invertebrate parasites came to take advantage of these predation events and became full-fledged parasites of the birds that ate them. Helminth parasites were not finished with invertebrates, however. In addition to using them as vehicles to get inside a bird, some present-day helminths (especially flukes) use invertebrates (especially snails) to reproduce asexually. In snails, flukes invade and consume tissues such as gonads and the hepatopancreas. Here, they undergo asexual reproduction, cloning themselves so that after being infected by a single fluke larva, an infected snail will end up releasing hundreds to thousands of flukes that can infect birds. This process greatly multiplies the local parasite population and floods the environment with propagules (offspring) that increase the odds that at least some of the parasites will make it to a bird. The process also means that flukes have evolved ways to bridge aquatic and terrestrial environments. Today, the invertebrate hosts are called intermediate hosts, while birds (the final bus stop on this life journey, where helminths become sexually mature) are called definitive hosts. Birds make great definitive hosts.

Once inside a bird's digestive system, worms find a variety of micro-habitats and a constantly replenished supply of high-calorie food. The digestive tract also provides an easy corridor that allows their eggs or embryos to exit the bird, so that the next generation of worms can eventually colonize another bird. For most helminths, the most dangerous time in the life cycle occurs when they are out of their bird home, exposed to the vagaries of the outside world.

GETTING HOME

Natural selection has forced helminths to perfect their abilities to get inside a bird—sometimes by using transmission methods that are so weird they defy belief. Researchers Bill Bethel and John Holmes discovered a bizarre method of worm transmission while sampling invertebrates in the shallows of some lakes near Edmonton, Alberta.[8] They noticed that freshwater amphipod crustaceans called scuds (*Gammarus lacustris*) would latch on—and cling tenaciously—to their hip waders and dip nets. Closer inspection showed that these scuds had bright orange spots just under the exoskeleton of their backs. In the lake, the orange-spotted scuds were found clinging to pieces of floating rushes, and when Bethel and Holmes brushed them off, the amphipods would skim along the water surface to find a bit of weed or other flotsam, and quickly grab on. If you put your arm or the back of a dip net in the water, the scuds with spots would cling.

Normal grayish-green-coloured scuds, without orange spots, did what scuds are supposed to do—scuttle along the top of the mud on the dark bottom of the lake, where they are camouflaged. When disturbed at the surface, normal amphipods quickly scatter away and dive to the bottom. Obviously, something here was strange.

In the lab, Bethel and Holmes found that the orange spots in the scuds were infective stages (called cystacanths) of an acanthocephalan worm, *Polymorphus paradoxus*. These worms were drastically disrupting the normal adaptive behaviour of the amphipods. Natural selection had, over millions of years, led scuds to spend their time on the bottoms of lakes, where they are

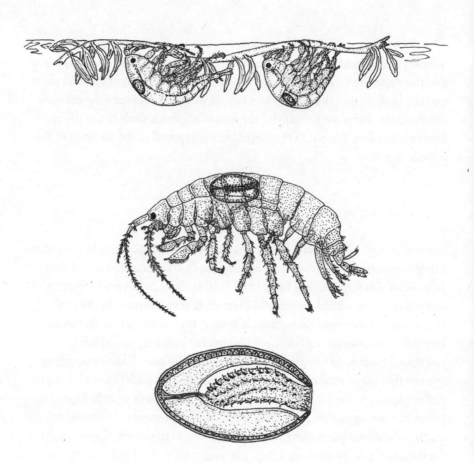

FIGURE 7.2 Two scuds, *Gammarus lacustris*, cling to vegetation at the water's surface. Their behaviour has been altered because they are infected with larval stages (cystacanths) of the acanthocephalan worm *Polymorphus paradoxus*.

concealed from daytime predators. But the scuds infected by acanthocephalan worms were "reprogrammed" to commit suicide by offering themselves as food for surface-feeding ducks such as mallards (*Anas platyrhynchos*). Bethel and Holmes noticed that infected scuds could even be seen clinging onto the feathers of their duck predators. And, as if luring the scuds to the surface weren't enough, the worms garishly advertised their duped amphipods' willingness to become food by flagrantly marking them with orange spots!

As unbelievable as this seems, the parasites were making amphipods change their normal behaviour to facilitate transmission to their bird host—a phenomenon that Bethel and Holmes termed "altered behaviour." In the lab, controlled experiments showed that infected scuds were attracted to light, clung to debris at the water's surface, and most tellingly, were eaten by mallard ducks four times more often than uninfected amphipods.

In later research, a student of John Holmes, Simone Helluy, learned how the cystacanths turn scuds into sacrificial zombies.[9] Like most crustaceans, amphipods have compound eyes that are composed of many facets. Helluy discovered that worms were likely secreting serotonin—an amphipod neurotransmitter—or inducing the scuds to secrete more of their own serotonin. This was causing melanin pigment in the compound eyes to be distributed abnormally, so that infected scuds at the water's surface perceived that they were safe at the bottom. The clinging response induced by parasites turned out to be behaviour that is normal during amphipod copulation—but the worms were able to hijack this behaviour and make scuds engage in it at times and in places that were in the worms' best interests, rather than in the scuds'. Consequently, infected scuds ended up being eaten by the usual worm host, mallard ducks.

Interestingly, another closely related spiny-headed worm, *Polymorphus marilis*, also infects scuds, but only affects their response to light and not their escape response. Thus, scuds infected with *P. marilis* get eaten by ducks like lesser scaup (*Athya affinis*).[10] Lesser scaup are diving ducks that feed in open water, so it appears that acanthocephalan worms alter their host's behaviour specifically to force scuds into the feeding niche where they are most likely to end up inside the parasites' favourite host—that is, the one where they will produce the most offspring.

Once research of this kind began to emerge, scientists started to notice many more examples of parasites changing the behaviours of their intermediate hosts to increase their chances of infecting their preferred definitive host.[11] Parasitic worms use other methods to get themselves into birds. These include weakening the stamina of their intermediate host, making their host more conspicuous, and disorienting the host.[12] Unfortunately for some intermediate hosts, helminths can employ more than one of these methods at the same time. For example, small fish such as spottail shiners, fathead minnows, pearl dace, and sticklebacks can be infected by tapeworm larvae (called plerocercoids) of *Ligula intestinalis* and *Schistocephalus solidus*. Worms are found packed in the abdominal cavities of fish. They can grow to 8 cm in length, which may be as long as the body of the host, and can account for 50% to 80% of the host's body weight. To put this in perspective, if you weigh 150 pounds (68 kg), then 75 to 120 pounds (34 to 54 kg) of your weight would be worm tissue.

Various studies have found that fish infected with plerocercoids are sluggish in their movements and lag behind their schools. Infected fish cannot swim continuously, and they use more oxygen than uninfected schoolmates. They swim at the surface in shallow, warm water near the shore.[13] Most fish are counter-shaded, with dark colouring on top and white bellies on the bottom, so that they are concealed from predators above or below them. However, fish whose bellies are swollen with plerocercoids can easily be spotted from above, because their pronounced abdomens give the appearance of two white stripes.[14] In some lakes, I have been able to catch infected fish by hand, and have had the unpleasant experience of having the tapeworm larvae burst out of the unfortunate fish's thin abdominal wall, right into my hand.

As you may have guessed by now, the definitive hosts of these tapeworms are various types of fish-eating birds, such as kingfishers, loons, grebes, gulls, and cormorants. In a study in Europe, it was found that infected roach (*Rutilus rutilus*) were almost five times more likely to be eaten by cormorants (*Phalacrocorax carbo*) than were non-infected fish.[15]

One wonders if birds are so stupid that they cannot learn to recognize and avoid worm-infected fish and thus avoid becoming the worms' permanent residence. John Holmes has suggested that the caloric benefits of easy prey (the equivalent of fast food for birds) far outweigh the energetic costs to birds associated with hosting these worms.[16] Thus, natural selection has not led birds to avoid infected intermediate hosts—in fact, it might instead prompt birds to prefer to eat them. If true, this is good news for both worms

FIGURE 7.3   A school of fathead minnows (*Pimephales promelas*) hunted by a red-necked grebe (*Podiceps grisegena*). One minnow is infected with the plerocercoid larval stage of the tapeworm, *Ligula intestinalis*. The worm has caused the infected minnow to swim behind the school and has affected its counter-shaded colouring, making the minnow more susceptible to predation. The worm will mature in a variety of fish-eating birds.

and birds—but I know of no studies that have been able to measure the energy costs and benefits to all partners. The kinds of tapeworms that get into birds this way mature very rapidly and do not cause any obvious pathology, so this may well be a win–win situation for these worms and birds.

After successfully entering the digestive system of the definitive host, helminths have a variety of microhabitats available to them. The gastrointestinal (GI) tracts of birds can vary greatly in structure, depending on the specific diet of the bird (fruits, seeds, insects, fish, carrion, and so on). Because food follows a strict sequence of mechanical and chemical steps as it passes through the GI tract and is broken down, each region of the digestive tract has become highly specialized. The first, or anterior-most, part of the GI tract includes the oral cavity and the esophagus. Birds have no teeth and swallow large pieces of food at a time, so helminths that inhabit birds are not subjected to the slicing and grinding that confronts worms that reside in mammal hosts.

The muscular esophagus forcefully pushes food downward. At its end, the bird esophagus often swells to form an expandable storage sac called the crop. Crops can be simple, or in some birds such as pigeons, they may be glandular to produce "pigeon milk," which is regurgitated to feed chicks. Some larval parasites can be transmitted from parents to young in this way.

A valve leads from the crop into a two-chambered stomach. The first chamber, the proventriculus, secretes acid and enzymes to break down food. The second chamber is the gizzard. The gizzard is a functional replacement for teeth—it has thick, muscular, rubbery walls that are covered by a gritty, sandpaper-like lining. In addition, it may contain quartz grains and small gravel that aid in macerating food. Grebes, which feed on fish and aquatic insects, have wads of their own feathers in their gizzards. Perhaps the feathers trap fish bones and insect exoskeleton parts, delaying their passage until they can be dissolved.

Few helminths have become specialized for life in the anterior part of the bird gut; those that infect sections beyond this point, however, must have

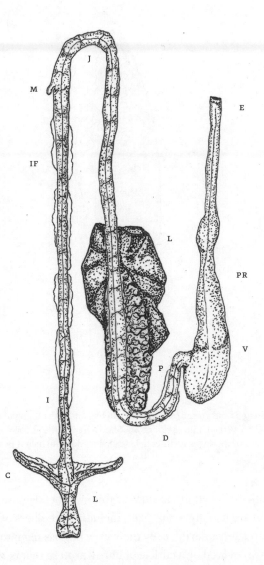

FIGURE 7.4   The digestive system of a bird, here, a fish-eating bird such as a grebe.
(E) Esophagus, the lower part of which includes an expandable pouch, the crop.
(PR) Proventriculus. (V) Ventriculus or gizzard. (D) Duodenum, the first part of the small
intestine. (P) Pancreas. (L) Liver. (J) Jejunum, the second part of the small intestine.
(M) Maekel's Diverticulum, a pouch of lymphoidal tissue. (IF) Intestinal fat. (I) Ileum, the
last part of the small intestine. (C) Caecum. (L) Large intestine.

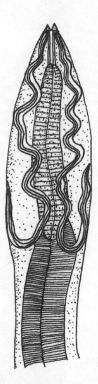

FIGURE 7.5 Some roundworms (nematodes) can live under the lining of the gizzards of birds where they may cause the koilin to slough off. Shown here is the anterior end of *Acuaria*, which has spines and elaborate ridges on its cuticle (called cordons) to help it pry under the lining.

adaptations that allow them to survive this gauntlet. Helminths that do live in the foregut are usually nematodes, including members of the families Acuariidae and Tetrameriidae. By their very nature, nematodes are tough animals. They are covered by a cuticle of horny protein that is often tanned like leather, so they can resist the mechanical and chemical assaults of the foregut.

Species of *Tetrameres* and *Microtetrameres* live in the stomachs of many birds. They are small (about 6 mm long), with males having roundworm-shaped cylindrical bodies. Females are swollen globs that are bright red in colour (perhaps due to ingested blood), and they live in the glands of the walls of the proventriculus and gizzard. Parasitologists collect them by using tweezers to pop them out of their homes.

Acuariids live protected lives under the gizzard lining (called the koilin). They are beautifully ornamented with cord-like expansions of the cuticle that extend from the head backward along their necks. In addition to these "cordons," spines and serrated flanges can also occur. These structures probably help the worms stay in place, resisting the strong contractions that characterize their home. Most gizzard worms occur in small populations, and they don't cause any problems for their hosts; however, the members of two genera, *Amidiostomum*, and especially *Epomidiostomum*, can cause the koilin of waterfowl to become ulcerated, inflamed, and eventually, to slough off.

Food from the stomach passes downward into the small intestine. The first part of the small intestine, the duodenum, is a descending loop that enfolds the pancreas. Digestive enzymes and bile salts (detergents that emulsify fats) enter the duodenum through pancreatic and bile ducts. Beyond the duodenum, the small intestine continues as the jejunum and ileum, where most of the nutrients from the bird's food are absorbed by finger-like intestinal villi.

The small intestines of different types of birds vary in length, depending on the type of food eaten. Birds that are largely carnivorous, such as raptors, tend to have shorter intestines compared to those that consume plant matter, such as geese.

At the junction of the small and large intestines are two blind pouches called caeca. The caeca are full of bacteria that probably help birds to further digest difficult materials such as plant cellulose. Again, birds with diets that include lots of plant matter have longer caeca. In most birds, the large intestine tends to be a short, straight tube that empties into the cloaca, a chamber that collects both the feces and urine.

All regions of the intestinal tract have a unique architecture. The thickness of the muscular walls, their degree of vascularization by blood vessels, the thickness of the digestive system lining (called the mucosa), the length of the villi, the depth of folds, and the number of mucus-secreting cells—all of these are features that vary from one portion of the tract to another. This creates many potential microhabitats that helminths can use as residences.

They can subdivide the intestinal tract and have developed specialized features to do so.

This is reminiscent of the way feather mites partition the plumage, as we saw in Chapter 5. There, we saw that air flow direction and velocity affect the body structures and distributions of feather mites. In Chapters 2 and 5, we saw that feather type, colour, and exposure to preening affect the distributions of both lice and mites. Similarly, worms sort themselves into distinct niches. Differences in intestinal architecture and, more broadly, regional differences in the mechanical and chemical aspects of the digestive system, create different niches for helminths.

MANY ROOMS, MANY OCCUPANTS

If we typically find that specific helminths stick to their own "rooms" within the host bird rather than spreading out throughout the digestive tract, what is it that drives them to these locations? Are they attracted to these places because these are the most comfortable locations for particular species? Or have they been forced to live there, pushed out of more desirable rooms by other helminths? Are some sites in the gut choice locations that worms must compete for?

These questions were addressed in a series of studies looking at the distributions of worms in the intestines of grebes (Podicipedidae) and waterfowl—such as lesser scaup ducks (*Athya affinis*) and white-winged scoters (*Melanitta fusca*)—that co-occur on lakes in Alberta, Canada. Researchers who study helminths in birds are confronted with certain difficulties that are not an issue for scientists who study ectoparasites. Ectoparasites can conveniently be observed on the outside of their hosts, but getting a good look at helminth populations requires destroying and dissecting the host bird. To document worm populations in these birds, the researchers had to collect and necropsy the birds in the wild, and, in order to preserve helminths *in situ*, flash freeze the birds' digestive systems. Later in the lab, the specific locations of worms could be determined. This method has inherent limitations—it allows researchers to get a snapshot of the helminth populations at a specific point in time, but it doesn't allow them to manipulate populations and study patterns over time, as is possible in research on ectoparasites.

Nonetheless, these studies of Alberta birds have yielded a rich picture of the helminth faunas in the intestines of lesser scaup, white-winged scoters, and grebes. The guts of these birds contain an abundance of worms from a rich variety of species, and each species tends to have a predictable location.[17] For example, Eric Butterworth and John Holmes noticed that a tapeworm (*Diorchis*) in the intestine of canvasback ducks (*Aythya valisineria*) anchored itself in the mucosa at a specific location, close to the place where the intestinal mesentery attaches to the outside of the intestine.[18] (The intestinal mesentery is the main structure that attaches the intestinal tract of a bird to the abdominal wall.) They also saw two species of small tapeworms in white-winged scoters that occupied the same area along the length of the intestine but embedded themselves into the mucosa at different depths. This suggests that helminths not only sort themselves in terms of where they live along the length of the intestine, but may be partitioning the zoo in other ways as well.

The characteristics of helminths vary dramatically and reflect their usual habitat within the bird. Intestinal worms vary in their feeding methods and can be sorted into three different guilds of worms.[19] Flukes and nematodes have mouths and digestive tracts, and thus belong to the engulfer guild (because they engulf host tissues or host food). Tapeworms and acanthocephalans, however, have no mouths or digestive systems; they are gutless wonders that absorb nutrients through their body surfaces. They are thus assigned to the absorber group, which actually contains two guilds. Some absorber helminths are small (1 to 2 mm) and attach and live within the confines of the host's mucosal surface (also called the paramucosal zone), while others are large (20 to 30 cm) and attach to the bird's mucosal surface but have the bulk of their bodies within the gut tube, or lumen. Thus, tapeworms and acanthocephalans could belong to either the paramucosal absorber guild or to the lumen absorber guild.

The theory of habitat selection suggests that organisms exist in areas where their per capita fitness is best—helminths living in the intestines of birds should be no different than other kinds of organisms in this regard. Because very different physical and chemical conditions occur along the intestinal gradient, worms have likely been moulded by different selection pressures.

The first third of the intestine of birds is subjected to pulsations of the villi, violent flow of semi-digested food, and the strongest contractions and peristaltic waves in the entire GI tract. This is a very active habitat. Worms living here can try to avoid being flushed out by being small enough to hide within the villi and folds of the mucosa (like an ocean barnacle wedging itself within the cracks of a rock that is being pounded by waves), or they may use well-developed holdfast structures to anchor themselves to the lining of the gut (like a mussel using its byssal threads to glue itself onto a rock's surface).

Tapeworms, which usually account for most of the helminths in the intestines of waterfowl and grebes, have a special holdfast organ at their head end called the scolex. This structure varies in different species of tapeworms and may include muscular suckers, retractable hooks, spines, tentacles, slit-like grooves, sticky glands, or various combinations of these. The scolices of tapeworms commonly found in grebes include four suction-cup suckers at the base of a retractable protrusion (called the rostellum) that is armed with one or more rings of sharp, chitinous hooks that vary in number, size, and shape, depending on the species of worm. One grebe-dwelling tapeworm (*Dioecocestus asper*) starts its life in the intestine as a young worm with hooks but loses these hooks as it grows and matures. As an adult, it crams its massive bulk of more than thirteen grams within the lumen of the gut and can avoid being dislodged due to its size. Others (*Ligula intestinalis* and *Schistocephalus solidus*) never have hooks or suckers but instead have two slit-like structures that are less effective for holding on— these helminths have only a brief time (seven to fourteen days) to live inside birds before being passed out.

Researchers discovered that, among tapeworms found in grebes, the length of their scolex hooks became shorter the further along in the intestine the worms took up residence—those found at the front part of the duodenum, where it is hardest to hold on, had the longest hooks. One species,

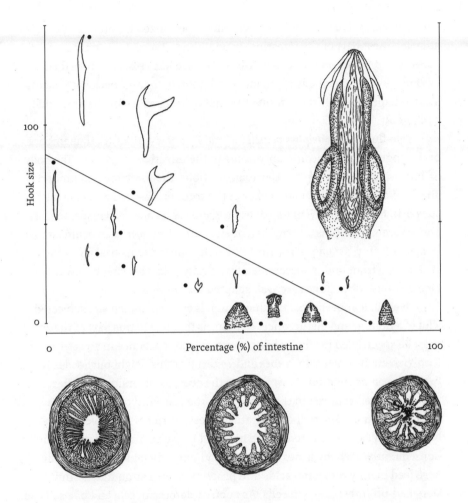

Hook size

100

0

0   Percentage (%) of intestine   100

FIGURE 7.6   The tapeworms of grebes that live at the anterior end of the intestine tend to have longer hooks and more elaborate holdfast structures than those living at the posterior end. This graph shows where tapeworms (Cestodes) of grebes (Podicipedidae) are normally found living along the length of the intestine in relation to sizes of their rostellar hooks (in micrometres). Tapeworms living toward the posterior part of the intestine have no suckers or hooks. A large tapeworm (*Dioecocestus asper*) living in the middle of the intestine has rostellar hooks but loses them as it grows to adult size. The architecture of the intestine varies along its length, as shown below the graph—at the anterior end, the muscular layer is thick and the mucosa has long finger-like villi. At the posterior end, the muscular layer is not as thick, the villi are more poorly developed, and indigestible debris tends to accumulate.

Pararetinometra lateralacantha, is extremely specialized for attaching to this anterior-most part of the intestine. In addition to the usual complement of suckers and hooks on the scolex, this worm also has crescent-shaped ridges on both sides of each body segment bearing thirty to forty backward-facing, sickle-shaped hooks! One can surmise that holding on in this habitat must be difficult.

Two flukes, *Tylodelphys podicipina* and *Petasiger nitidus* (both members of the engulfer guild), commonly occur in the intestines of grebes. The body of *Tylodelphys* is divided into two main sections, the forebody and hindbody. The forebody has a muscular sucker at the front end as well as a ventral sucker in the middle. With its adhesive glands and cone-like shape, the entire forebody acts as an attachment structure. *Petasiger* is an echinostome (*echino*—"spiny"; *stoma*—"mouth") fluke. In addition to its oral and ventral suckers, its front end is surrounded by a fleshy collar bearing backward-facing spines. Both flukes are well adapted for holding on.

Overall, it appears that the duodenum is a nutrient-rich site subjected to a lot of agitation and disturbance. Helminths that live in this part of the zoo must be specialized for their ability to attach and reproduce in an active but food-rich environment. In grebes and waterfowl, this neighbourhood is home to the largest populations of worms and the greatest diversity of species.

In marked contrast to the duodenum, the last third of the intestine (the ileum) is a harsh, desert-like habitat. The best nutrients—simple sugars—have already been extracted by the host and by parasites living upstream. Consequently, carbohydrates are scarce and unpredictable. Worms living here feed mainly on amino acids and proteins as well as on mucus and sloughed-off intestinal skin cells. As another deterrent, bile and bile salts are released in the duodenum and, in the ileum, act as fat emulsifiers—essentially, as detergents.

Although peristaltic waves are much weaker in the last part of the intestine than in the rest of the gut, resident worms are housed in a detergent bath with small amounts of low-quality food. Coarse, indigestible debris accumulates. I imagine that for birds that eat a lot of plant matter, toxic alkaloids and other chemicals probably gather here. This part of the gut is subjected to abrasion, so the mucosal surface is constantly being sloughed.

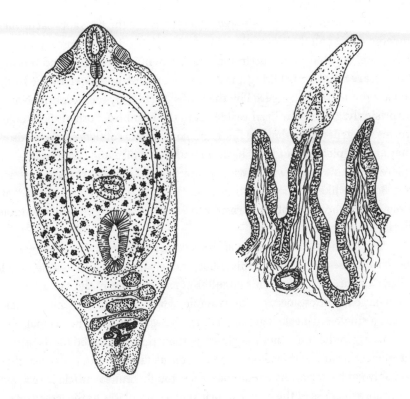

FIGURE 7.7   A fluke (*Tylodelphys podicipina*) has muscular suckers, adhesive glands, and a cup-shaped body that it uses to attach and feed in the intestine of grebes.

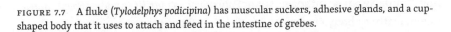

This area, along with the large intestine and cloaca, is where much of the water used in digestion is resorbed, so it is much drier here than in the rest of the GI tract. For worms, this is skid row.

Worms living here make use of at least two general strategies for survival. They can cope with the lack of food by using nutrients they have acquired and stored from their previous life in an intermediate host and limit their time spent in the bird. Or, like kangaroo rats in a desert milieu, they can evolve specialized adaptations that allow them to spend long periods of time in their harsh environment. Both strategies occur. Tapeworms like *Ligula intestinalis* and *Schistocephalus solidus* use the first approach. They undergo almost all their growth as larvae in fish. As a result, once in the flying zoo, they have few nutritional requirements. Maturation and reproduction are rapid (only taking about 24 hours after being eaten by the bird) and are immediately triggered by the bird's warm body temperature. Not surprisingly, given their brief residence in the host bird, both of these tapeworms are generalists that are found in many species of fish-eating birds. They are not picky about their location within the bird and can live anywhere from 20% of the way down the intestine to all the way to the posterior end. Both are likely opportunistic worms that take advantage of any vacant room (or are forced by other parasites to use it), including the unfavourable distal part of the gut.

One species that makes use of the second strategy, having developed special adaptations for life in this inhospitable part of the zoo, is a nematode called *Capillaria obsignata*. This engulfer normally hides in the caeca, but when large infections occur—for example, when thirty to forty nematodes invade a single red-necked grebe (*Podiceps grisegena*)—some individuals spill over into the intestine. They burrow their thin, elongated heads into the mucosal lining and feed on blood and tissue (and, within the caeca, they probably eat bacteria). As nematodes, their tough, tanned cuticle is resistant to the abrasiveness of the coarse debris that accumulates in the lower intestine. Similarly, an acanthocephalan, *Polymorphus marilis* (an absorber), has a tough, leathery body surface and can endure here too.

The middle third of the intestine (the jejunum) is a compromise between the conditions of the anterior and posterior parts. This area is characterized by high and predictable concentrations of prime-quality nutrients, lots of water, and moderate levels of disturbance from peristaltic contractions. The villi and intestinal folds in this zone are long, luxuriant, and well developed, and they provide good habitat for worms. For helminths, this is prime real estate in the flying zoo. You might expect the helminth population density to be higher here than in the more turbulent anterior regions of the intestine. However, this is not necessarily the case. In red-necked grebes, portions of the mid-intestine are often hogged by a single species of massive tapeworm known as *Dioecocestus asper*.

Dioecocestus asper is a large, lumen-dwelling absorber that makes its home in the prime location of the mid-intestine. While most tapeworms are simultaneous hermaphrodites (that is, a single worm has both functional male and female reproductive systems), *Dioecocestus* has separate male and female individuals, which grow to be very large—as mentioned earlier, large enough that they no longer need hooks as adults, relying on their size to stay wedged within the lumen. Stringent population regulation occurs—of 27 infected red-necked grebes, 24 had a single mixed-sex pair of worms. How this happens is a mystery, but when single worms are found, they are always female, so perhaps some system of conspecific sex determination occurs, where the first animal in a habitat develops as one sex, then releases pheromones that cause any other animals of the same species to develop as the other sex. (This process is also known for some deep-sea fish.)

Besides its large biomass and constant population size, *D. asper* has a low reproductive rate—this is a very unusual trait for tapeworms, which are notorious for producing thousands to hundreds of thousands of eggs a day. Moreover, *Dioecocestus* has a long generation time of at least a year, in contrast with several weeks for most tapeworms found in waterfowl.

The giant size of this worm causes damage to the mucosal surface of the mid-intestine of grebes. Where it lives, the villi are flattened and eroded—probably due to necrosis resulting from pressure. Also, the worms create intestinal blockages, so that upstream of *Dioecocestus*, plugs of green, bile-stained luminal contents and mucus can be found. The mid-intestine containing *D. asper* is stretched to more than 150% of its normal diameter, to the point where the intestinal wall is so thin that the worms are visible from outside the intestine.

When researchers removed *Dioecocestus* worms during necropsy, the mucosa had tire-track patterns corresponding to the imprints of the individual segments of the tapeworm. Grebes may try to compensate for the damage by increasing the number and size of villi in the intestine upstream of *D. asper*. In any case, fortunately for the host, only a relatively small part of the intestine is affected.

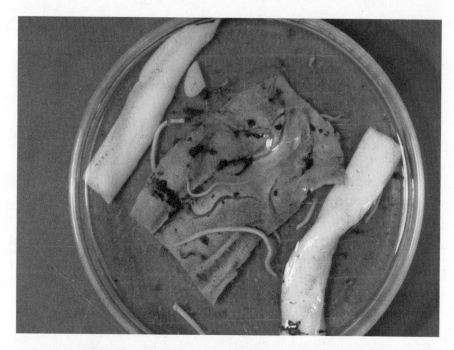

FIGURE 7.8 An unusual tapeworm, *Dioecocestus asper*, lives in the intestines of red-necked grebes (*Podiceps grisegena*). It uses its bulk to wedge itself in the intestine and to drive out competitors, except for a tough, large nematode (*Contracaecum ovale*), which lives in the same area of the intestine. *Dioecocestus* creates pressure on the intestine, causing the villi to be short and flat and *Contrcaecum* to become pressed into the mucosa.

In addition to its effects on the host, this worm influences other helminths, driving them out of the prime mid-intestine locale.[20] Worms normally found upstream of *D. asper* shift their positions further upstream, toward the duodenum, with those that normally occur closest to *Dioecocestus* shifting the most. The presence of this giant tapeworm has been linked to reduced population sizes of other worms in the mid-intestine, and it has been associated with a reduction of 22% in the number of species of other helminths in red-necked grebes.

Only one helminth species seems unaffected by *D. asper*—a large nematode engulfer called *Contracaecum ovale*. Its distribution in the intestine of red-necked grebes is virtually identical to that of *D. asper*. In fact, it seems to prefer to live in the same part of the intestine, even though (or perhaps because), it gets pressed into the mucosal surface by its giant roommate. *Contracaecum* is a robust nematode. Perhaps it feeds on the disrupted host tissues or on the food blockage created by *Dioecocestus*. When it occurs in birds without *D. asper*, *Contracaecum* lives in the same area of the mid-intestine and does not cause any noticeable intestinal pathology.

Overall, *Dioecocestus* is a serious competitor in the worm community in red-necked grebes. It is a bully that uses its bulk to exclude other helminths from the choicest location in the intestine. If you were to compare its mass to that of other grebe parasites, *Dioecocestus*'s presence in the intestine would be like having an African elephant living with a small dog in a dark, tubular enclosure barely big enough to house the elephant. Nevertheless, because its population size is regulated, *D. asper* seems to show a great degree of co-evolution with its normal host, the red-necked grebe. It is likely a keystone species in the grebe parasite community because it affects the numbers, types, and locations of other worms.

PATTERNS OF CO-EVOLUTION

If the habitat characteristics of the intestines of grebes are like those of other birds (or even other vertebrates), then some general trends in the communities of helminths can be predicted, as suggested in Table 7.1. However, the patterns seen in grebes may not be widely shared with other animals. There are two main reasons for this. First, grebes have a broad diet of fish and aquatic invertebrates. This ensures that grebes are repeatedly exposed to many different worms. Second, grebes (family Podicipedidae) are an ancient group of birds. Fossils ascribable to the extant genera *Podiceps* and *Podilymbus* are known from the Miocene some 20 million years ago.[21] Thus, grebes have acquired a very specialized—and probably co-evolved—suite of helminth parasites. In younger communities, the ecological patterns seen in grebes may not have yet had time to form.

TABLE 7.1.   Predicted habitats and helminth parasites in the intestines of birds

| | REGION OF INTESTINE | | |
	ANTERIOR (DUODENUM)	MIDDLE (JEJUNUM)	POSTERIOR (ILEUM)
HABITATS			
Nutrients	Rich and plentiful	Rich and plentiful	Poor and scarce
Water available	Plentiful	Plentiful	Reduced
Coarse debris/toxins	Moderate/moderate	Moderate/low	High/concentrated
Disturbance	High	Moderate	Low
Complexity	Complex with long villi, deep crypts	Moderate	Simple with short villi, shallow crypts
HELMINTHS			
Biomass	Moderate	High	Low
Number of species	High	Low	Low
Population size	High	Low	Low
Average body size	Small	Large	Variable
Vacant space	Moderate	Rare	Common
Strategy	Competitive, ruderal*	Competitive	Stress tolerant
Competitive style**	Exploitation	Interference	Avoidance
Host specificity	Narrow	Moderate	Wide and narrow
Reproduction	Rapid	Slow	Rapid and slow
Dominant guild	Engulfers and paramucosal absorbers	Luminal absorbers	Engulfers
Degree of co-evolution	Moderate	High	Low

* Ruderal—like a weed; can rapidly colonize disturbed sites.
** Exploiters try to use resources before others can get access; interferers block others from resources.

As we have seen, helminths need a variety of adaptations to colonize, survive, and reproduce in different hosts. If parasites co-occur in birds regularly and predictably, then there is the potential for co-evolution, not only between the parasites and their hosts, but also among the various kinds of parasites within a host. A complex collection of worms that have been together in the same host for a long time (we might call them the core species) may even benefit their host by protecting it from more dangerous parasites.

There is some evidence that this may be the case for wild mute swans (*Cygnus olor*), which, like grebes, have a mature community of helminths. However, in a zoo environment, the swans are not exposed to their usual parasites. Mute swans in a zoo were observed to become infected with two tapeworms (*Dicranotaenia coronula* and *Diorchis stefanskii*) that are normally rare and do not mature in swans.[22] Furthermore, the swans in the zoo, lacking their normal worms, acquired large infections of abnormal parasites that caused disease. It seems that normal worms in mute swans have established a well-balanced relationship with their typical host and with each other, and all lead lives of peaceful co-existence. By protecting their hosts from pathogenic parasites, they may be considered mutualists.

Co-evolution between hosts and helminth parasites is not a new idea. It has been around for a long time and has led to the proposal of several rules,[23] as discussed in Chapter 2 for lice. For example, Fahrenholz's Rule claims that parasite evolutionary histories, or phylogenies, should mirror the histories of their hosts. A corollary of this idea would be that a host that shares a parasite may be evolutionarily related to other birds that are infected by the same worm. Another rule is Manter's Rule, which says that long associations between hosts and parasites should lead to strong host specificity.

The intestinal helminth communities in western grebes (*Aechmophorous occidentalis*), red-necked grebes (*Podiceps grisegena*), and eared grebes (*P. nigricollis*) are each composed of a group of parasites that frequently and consistently co-occur (the core parasites), in addition to a group of parasites that occur sporadically (the satellite parasites).[24] The presence of core parasites in each type of grebe means that each species has a characteristic and predictable helminth community—even though many of the core parasites are shared by more than one type of grebe.

One family of specialized tapeworms included among the core species of grebes is the Amabiliidae family. It includes species in two genera, *Tatria* and *Schistotaenia*. Amabiliids (with one important exception) are host-specific to grebes. It turns out that comparing the evolutionary histories of *Tatria* and *Schistotaenia* with that of grebes, to see if they follow the co-evolutionary rules, is fraught with problems—not the least of which is determining which host grebe is the typical host.

Often, several species of grebes nest on the same lake, and although they do have distinctive diets, food overlaps among them are common.[25] Hence, helminths are shared among grebes. When quantitative data are available, the main hosts can be determined, but for most of the world's grebes, the parasite communities have not been closely examined or enumerated. This makes it difficult to determine whether there has been co-evolution (including co-speciation—the formation of a new species of worm each time a new species of host forms). Sometimes, researchers must rely on zoo-geographic patterns for help.

Nevertheless, a first attempt at matching amabiliid tapeworm and grebe phylogenies showed alignments in both zoo-geographic and co-speciation patterns. For *Tatria*, five species of worms seemed to show co-speciation patterns, but two cases of colonization, or host capture, where parasites infect hosts not recently related to each other, were also indicated. One species of worm, *Tatria biremis*, evolved from an Old World ancestor that infected a grebe in the *Tachybaptus* lineage (for example, the European little grebe—*T. ruficollis*). The main host today is the Old World eared grebe (*Podiceps nigricollis*). Two Polish researchers, Wanda Korpaczewska and Teresa Sulgostowska, recognized two races (subspecies) of *T. biremis*— *T.b. minor* was specialized in eared grebes and *T.b. major* occurred in red-necked, eared, and great-crested grebes (*Podiceps cristatus*), as well as in little grebes.[26] If one of the races of the worm has become specialized for eared grebes, this suggests an intimate evolutionary relationship between the two, and provides support for eared grebes as the main host.

The patterns of host capture and co-speciation, along with their relationship to zoo-geographic patterns, point to a long relationship between amabiliid tapeworms and grebes. The genealogical comparisons of both *Tatria* and *Schistotaenia* with grebes showed that the divergence between Old World and New World hosts, which probably occurred in the mid- to late Cretaceous (some 80 million years ago), was an important co-evolutionary event. Most evidence suggests that grebes have lived with a suite of parasitic worms (especially amabiliid tapeworms) for a long time. However, there is one species of amabiliid tapeworm that does not occur in grebes. It is found in flamingos.

Flamingos are well-known and well-loved birds. Often, these long-legged, pink birds are considered comic (the plastic variety being used to embarrass someone who has reached a milestone birthday), or they are thought of as stylish and stately. As an undergraduate student doing a lab in animal behaviour, I once spent three hours patiently observing a small flock of flamingos in a zoo. During that time, the most exciting behaviour they demonstrated was altering which single leg they balanced on while standing in a pool with their heads tucked under their wings. Nevertheless, they do have many interesting and special features.

Their "upside down" mouths allow them to strain invertebrates and algae from saline lakes and lagoons throughout the tropics of Africa, southern Eurasia, the Galapagos Islands, the Caribbean, Florida, South America, and on some lakes high in the Andes Mountains. As discussed in Chapter 2, their evolutionary history and the identity of their closest relatives has been a great enigma in biology.

Scientists have identified 30-million-year-old fossils of the modern flamingo species—and specimens of more primitive types date to 50 million years ago. Because of their long-legged, wading lifestyle, flamingos have been identified as genetically close to herons and storks (order Ciconiiformes). This interpretation was supported by a study in which DNA molecules were separated into single strands and allowed to recombine with other DNA molecules.[27] (The idea behind this technique is that the more recently related two organisms are, the more similar are the sequences of DNA base pairs. When double-stranded DNA molecules from two organisms are separated into single strands and mixed together, they will combine. The more base pairs in common, the more such hybrid molecules will form.)

However, this remained a controversial interpretation. Flamingos have unique bills with long, comb-like structures (called lamellae) used for filter feeding. They also have webbed feet, they vocalize by honking, and they have bile acids that are similar to those of geese (order Anseriformes), suggesting a possible relationship to Anseriformes. To make matters even more complicated, Storrs Olsen and Alan Feduccia thought that flamingos were closely related to stilts and avocets (family Recurvirostridae), and suggested that they be put in the order Charadriiformes.[28] As a solution to this problem, the five species of flamingos were placed in their own order, the Phoenicopteriformes—but no one could agree if they were related to herons, geese, or stilts.

These arguments were turned upside down when three studies were published, based on molecular evidence along with evidence from bodily characteristics, indicating that the closest relatives of flamingos are grebes![29] This evidence suggests that flamingos and grebes share a common ancestor. It is hard to imagine two groups of birds that, superficially, look more different. Flamingos are tall, stately, long-legged, web-footed waders, while grebes are divers, with short legs, placed so far at the backs of their bodies that they can barely move on land. Grebes have specialized feet with flanged toes that are rotated in water (like propellers), so they can blast along underwater, spearing fish and invertebrates with their harpoon-shaped bills. In fact, loons (order Gaviiformes), which are also diving birds, look very much like grebes, and yet they are distantly related to grebes—they look similar due to convergent evolution.

George Sangster has gone so far as to formally recognize the relationship between grebes and flamingos by creating a new taxonomic group for both, the order Mirandornithes (*miranda*—"wonderful"; *ornis*—"birds").[30] The idea that grebes and flamingos are sister groups was strongly contested when it was first proposed.[31] However, parasites seem to support the idea. Perhaps if we had more carefully considered the evidence provided by worms, the relationship between grebes and flamingos would not have come as such a surprise. Besides grebes, flamingos are the only other known host for a specialized amabiliid tapeworm belonging to the genus *Amabilia*. Most amabiliid tapeworms use dragonflies and damselflies as intermediate hosts—primitive aquatic insects that have existed for at least 300 million years.

Feather lice, which, like helminths, spend their adult lives with one host, should also indicate evolutionary affinities. As we saw in Chapter 2, a first look at louse populations in flamingos suggested that they were related to ducks, geese, and swans (order Anseriformes), because flamingos are infested by feather lice in the genera *Anaticola*, *Anatoecus*, and *Trinoton*, parasites found on Anseriformes, which seemed to support the idea that flamingos were related to waterfowl.[32] However, after the molecular evidence relating grebes to flamingos was published, the lice were reinterpreted by Kevin Johnson and his colleagues, who found good evidence suggesting that the specialized feather lice of grebes (*Aquanirmus*) are a sister group of the *Anaticola* lice of flamingos.[33] The researchers concluded that the close relationship between these lice was due to the fact that their hosts shared a common ancestor (that is, co-speciation had occurred). They also offered the surprising suggestion that the similar louse faunas of flamingos and ducks occurred because of host switching—from flamingos to waterfowl! In the end, parasitological evidence from both tapeworms and lice verified the idea that grebes and flamingos are related. The broader lesson is that if we learn more about the phylogenies of parasites, we will improve our understanding of the biology of their hosts—the residents can tell us a lot about the zoo!

UNDESERVED REPUTATIONS

Parasitic worms are usually regarded with disgust and contempt. We associate them with disease, poor living conditions, and shoddy hygiene. They have a sleazy history—at one time, tapeworm eggs were sold to people as a weight-loss remedy, the idea being that worms would steal so many calories from their host that the pounds would fall off. However, when we examine wild birds, our perception changes. Often it is the largest, healthiest-looking bird that harbours the largest population of intestinal helminths—which should not be surprising when we realize that transmission is usually caused by eating infected intermediate hosts in large quantities. Even worms that can potentially cause damage to their host (such as blood flukes in the circulatory system or a tapeworm like *Dieococestus* in the intestine) self-regulate their populations to avoid causing superinfections.

Blood flukes have a very interesting relationship with their hosts. Two major kinds exist—those that infect mammals (including humans) and those that infect birds. Both have identical life cycles that use aquatic snails as intermediate hosts, but instead of becoming infected by eating a fluke-infected snail, definitive hosts acquire flukes directly when free-living, actively swimming larvae (called cercariae) burrow through skin.

Once inside their host, worms travel in the bloodstream to the liver. Unlike most flukes, blood flukes occur as separate male and female worms. The worms pair in the liver. Male worms are larger and have a wide groove (called the gynocophoric canal) within which they embrace a slender female. The gynocophoric canal makes male worms look as though their bodies are split down the middle, giving blood flukes their scientific name, schisto-somes (*schisto*—"split"; *soma*—"body").

Female worms cannot mature sexually unless they are held by a male. Locked in embrace, the pair make their way from the liver, against the flow of blood in the hepatic portal veins, and wedge themselves into blood vessels as close to the intestine as they can get. One species in birds, *Trichobilharzia regenti*, uses the central nervous system as a migratory route and lives in the nasal passages. A male's body surface is covered by spines and bumps that, along with his oral and ventral suckers, help the amorous couple resist the blood flow and stay in place.

Although staying anchored in veins is a little tricky, the worms are constantly bathed in blood that is rich in dissolved nutrients, straight from their host's digestive system, so food is not a problem. Getting their eggs out into the world, however, is a challenge. A well-nourished female schisto-some lays several hundred eggs per day against the inside lining of a vein. Many eggs get swept away to their death in the liver, even though the eggs have spines that help them to stick. The spines on fluke eggs help them work their way through the blood vessel wall, and, unless trapped and killed by their host's immune system, to eventually break through the mucosa of the gut along with a bit of host blood. Later, the eggs will be expelled with their host's feces. In water, the eggs hatch to release a ciliated larva that finds and penetrates a snail.

In theory, damage caused by blood flukes in a bird or mammal can be great. Eggs that end up in the liver cause inflammation and are walled off in capsules and destroyed. Over time, the liver can become enlarged, with lots of non-functional scar tissue. The excretory products of adult worms are toxic and have to be cleared in the liver. The spleen can become enlarged to replace lost blood, and the host can become anemic. The most serious damage caused by blood flukes, however, results from the eggs. Most never make it out of the host but get trapped between the vein and the gut lumen. Like the eggs in the liver, these cause inflammation, become walled off by connective tissue, and over time, large parts of the functional digestive system become non-functional scar tissue. In humans, this can lead to cancerous tumors.

Nevertheless, although natural infections with blood flukes are common among ducks, geese, swans, and other waterbirds, and despite the damage flukes can potentially cause, disease as a result of infection is rare.[34] Evolution has led to a benign co-accommodation between worm and bird, taming the potential for schistosomes to kill their host. Adult schistosomes living in the veins provoke strong immune reactions in their hosts, but worms have evolved ways to avoid being destroyed. Frequently flukes change their surface coats and cover themselves in host molecules, which provides camouflage from the immune system.

In a bird infected with adult flukes, newly arrived larvae are readily attacked and killed. This protects their host from getting superinfected— only a few pairs of adult worms will be found in any bird, and the bird can live for years with its blood flukes. This process is called concomitant immunity.

Although you may not have been aware of bird blood flukes before, you may have unknowingly encountered them, especially if you've been swimming in lakes that have lots of vegetation such as reeds and bulrushes that are inhabited by snails; you may have come out of the water with a very irritating rash called swimmer's itch. The cause of this ailment turns out to be duck schistosome cercariae, which try to penetrate human skin. Fortunately for us, they do not survive inside.

When worms in the flying zoo do cause trouble, it is often indicative of some other underlying problem—for example, a host that is malnourished or has been exposed to toxic pollutants. Or it may be that sick hosts somehow facilitate transmission to other hosts. For example, some parasites may reproduce so rapidly in a host that they cause the host to be sick. In some cases, making their host weak or debilitated will actually increase their chances of being eaten by another host. Sometimes pathology caused by helminths points to a relatively new host–parasite association—as happens when a host expands its range, or when a parasite has been introduced into a naïve host population.

Worms may turn out to be very valuable to humanity. As we expand our knowledge of worms and of what they must do to survive in hosts like birds, we may be able to use this information to cure our own diseases. Worms, such as blood flukes, and others living in the lungs and intestines, have evolved many tricks to avoid being killed by the host's immune reactions. One method is to modulate, or down-regulate, the immune response, or to misdirect it to attack harmless targets. Some human diseases are caused by immune system attacks on our own tissues. These are called autoimmune diseases, and include rheumatoid arthritis, multiple sclerosis, type 1 diabetes, and inflammatory bowel disease. It is very likely that we will soon learn how helminths alter host immune systems; perhaps we will be able to use these methods to control, or even cure, autoimmune diseases.

Worm infections in wild birds are not unnatural. They are a normal and common part of the biological world. Every bird you see is carrying one or more helminth worms inside. The more we learn about helminths in birds, the more they tell us about evolution and ecology and about the intricacies, complexities, and beauty of life.

8

O D D I T I E S

I N T H E

F L Y I N G Z O O

IN THE MARGINAL ZONE

It seems that every zoological garden has its oddities—those few rare and bizarre animals that draw many people to see them. Often, these are animals that most of us have never heard of, let alone know where or how they live. The flying zoo has its own oddities, and even biologists who spend their professional lives working with bird parasites and diseases know that there are some kinds of animals that they will likely never encounter. In this chapter, I want to introduce you to some of these weird and wonderful members of the flying zoo.

Each of these animals has its own origin story, and its own reason for being relegated to the margins of the zoo. Some have evolved in isolation and exhibit peculiar and highly unusual habits that allow them to interact with their host birds. Others may have survival strategies that are very similar to the zoo's more common occupants, but for whatever reason, have not diversified in the same ways as their more successful relatives. Still others may be restricted to specific geographic regions, where they opportunistically make use of birds as hosts in addition to exploiting other animals within that ecosystem. Occasionally, some of these parasites can only be found associated with one kind of bird, so most of us would never guess they exist.

Together, these oddballs hint at the diversity of life forms and the myriad ways in which biological organisms attempt to get a foothold, sometimes flourishing, and sometimes hitting dead ends on their evolutionary tree.

TEAR-DRINKING MOTHS

Islands are wonderful places to look for oddities because their isolated geography offers many opportunities for animals to develop along unique evolutionary paths. And big islands may offer a bonanza of unusual species. Madagascar, an island nation in the Indian Ocean, is no exception. Although today the island is close to the east coast of Africa, geologists believe that it was once part of the Indian subcontinent. It is the fourth largest island on Earth—slightly smaller than the state of Texas. The island has fascinated biologists for a long time, because it contains about 5% of the entire world's species of plants and animals. About 80% of these organisms live nowhere else on Earth. These include the fossa (*Cryptoprocta ferox*), a carnivorous mammal that is probably related to the mongoose, as well as various species of lemurs—mysterious prosimians that are mostly nocturnal and live in the rainforest canopy. Five unique types of birds occur there as well, all with strange-sounding names: ground rollers, cuckoo rollers, asities (including sunbird asities), mesites, and vangas or sicklebills. The gigantic, flightless elephant bird (*Aepyornis maximus*), which weighed over five hundred kilograms and stood more than three metres tall, once ruled Madagascar, but was driven to extinction in historical times. Still, even today, Madagascar is a land full of wonderful and unusual animals.

During a biological survey to assess the biodiversity of the island, Roland Hilgartner and Mamisolo Raoilison observed some moths on the heads and necks of sleeping magpie robins (*Copsychus saularis*) and on

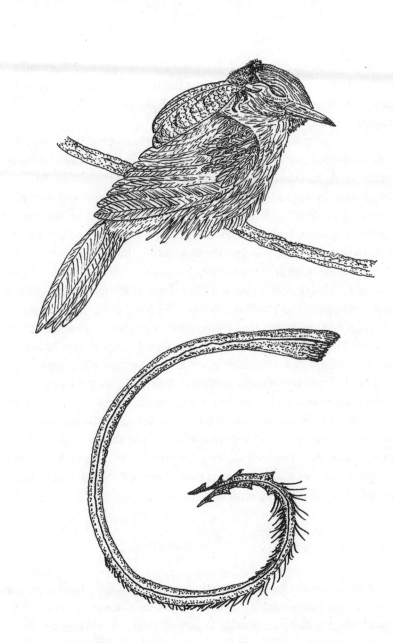

FIGURE 8.1 A moth (*Hemiceratoides hieroglyphica*) feeds on the tears of a sleeping Newtonia bird in Madagascar. The mouthparts of the moth have barbs and spines that penetrate under the eyelids of the birds.

several species of passerines called Newtonia birds (*Newtonia*).[1] The moths (*Hemiceratoides hieroglyphica*) were not just resting on the birds—they were inserting their long, coiled mouthparts through the eyelids of the birds and drinking their tears.

Tear-drinking moths had been observed previously, but these used mammals as hosts, especially ungulates. Animals that are attacked by moths are usually docile and have no paws with which to shoo them away. In the rainy season in tropical rainforests, salt becomes valuable and scarce, making tears a desirable source of salt. But on Madagascar, there are no large native ungulates, and the mammals that are there, such as lemurs and fossas, do have paws, motivating moths to look for other hosts. However, if you are a moth using birds as hosts, one challenge is that birds have two eyelids, both of which are tightly closed during sleep. Evolution has provided an ingenious solution to this problem. The tear-drinking moths of Madagascar have long, coiled mouthparts that act as straws to sip tears; these are armed with recurved hooks and barbs that can be inserted under the eyelids of their sleeping hosts. This adaptation is not known for any other tear-drinking moth. Since this strange moth was first discovered, many questions have been raised. Do the moths produce an anesthetic, to avoid waking up their hosts? Do the moths cause pathology in the birds? Do the moths also take blood, which is another source of salt (given that related species of moths are known to be blood-feeding vampires)? The unusual habits and body shape of this moth have intrigued researchers who would like to better understand their relationship with their hosts.

TONGUE WORMS

The flying zoo contains a group of animals that most zoologists have never even heard of, the class Pentastomida, "commonly" known as tongue worms. For a long time, biologists didn't know what to do with them: Should they be considered as unique animals and placed in their own phylum? Or are they aberrant types of helminth worms, somehow related to flukes or tapeworms, or perhaps some kind of annelid worm (related to leeches or marine rag worms)?

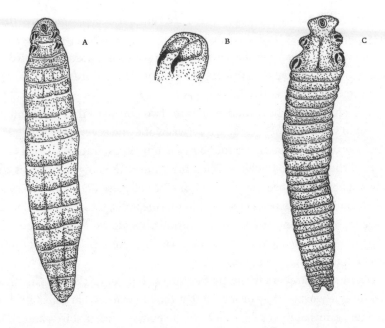

FIGURE 8.2   Tongue worms (phylum Arthropoda; class Pentastomida) that infect birds.
(A) *Hispania vulturis* from a black vulture. (B) Hooklets project from small appendages
(called parapodia). (C) *Reighardia sterna* from a gull.

The animals are legless, wormlike, and move sluggishly, like annelid
worms. Superficially, they seem to have segmented bodies, but their bodies
are covered in a cuticle that—like the exoskeletons of arthropods—contains
chitin. Tongue worms also resemble arthropods in that they have striated
muscles and two pairs of hollow, retractable hooklets near their mouths
that are rudimentary appendages. These features eventually led biologists
to conclude that tongue worms are members of the phylum Arthropoda—as
are insects, mites, and ticks. In fact, the immature stages of tongue worms
resemble mites.

Tongue worms evolved as endoparasites in the respiratory systems of
reptiles, at least since the Mesozoic.[2] Today, 90% occur in the lungs of croco-
dilians, monitor lizards, and constricting and venomous snakes. Modern

analysis indicates that tongue worms are very derived crustacean arthropods, perhaps most closely related to ectoparasites of fish (brachiurans). The tongue worms of reptiles use vertebrate intermediate hosts (fish, birds, and mammals), so perhaps the few tongue worms that mature in birds originally used them as intermediate hosts.

Tongue worms bear their names for two reasons. First, their bodies resemble small tongues (most are between one and several centimetres long). Second, they often are found in the upper respiratory system of their host. Of the 130 or so known species, most infect the lungs of reptiles, and only 3 species have been found in birds. Two of these (*Reighardia sternae* and *R. lomviae*) have been observed in marine gulls, terns, and guillemots. However, at least one other tongue worm (*Hispania vulturis*) has been found living in the trachea and air sacs of a very different bird, the black vulture (*Aegypius monachus*), in Spain.[3]

When researchers found that vultures, and not just marine birds, could host tongue worms, they proposed that tongue worms that infect birds can have two kinds of life cycles—one for the marine, coastal birds and another for carrion-feeding vultures. They suggested that gulls, terns, and guillemots get infected with tongue worms by eating fish that contain their larvae. The only hard evidence supporting this idea came from a study of common gulls (*Larus canus*) in Norway. There, males were known to eat more fish than female gulls. Male gulls were also found to be infected with tongue worms more often than females, and when they were, they had more worms than female gulls.[4] This supported the notion that tongue worms originated as larvae in fish. However, a curious observation, noted by several researchers, was not consistent with this idea: worm eggs and larvae in the primary, secondary, and older stages *all* occurred in the digestive systems of birds, with adult tongue worms occurring in the birds' lungs. This suggests that tongue worms have a direct life cycle, living out their entire lives in the bird, and not involving fish or marine invertebrate as intermediate hosts.

Gulls frequently congregate at the shoreline, and so accidental ingestion of feces, or of feathers contaminated with feces, may result in the transmission of tongue worms. The worms may also be transmitted when disturbed gulls regurgitate semi-digested food balls (called *speiballen*) that are hungrily devoured by other gulls. Gulls also feed chicks by regurgitation. A mucus-covered food ball regurgitated by an adult infected with tongue worms could contain eggs, which would in turn infect the chicks.

Apparently, tongue worm eggs only need body heat as a cue to hatch. According to A. Banaja and colleagues, the primary larvae use their hooklets to break out of eggs and then crawl through the wall of the host intestine.[5] This takes only one hour. In the bird's body cavity, larvae immediately moult. Within two hours, only second-stage larvae can be found. Then, for about two weeks, these second-stage larvae crawl over the surfaces of the intestines, at the end of which time they use their hooklets to enter abdominal air sacs. The larvae then migrate forward to the anterior air sacs, staying there for 10 to 20 days and become sexually mature. Here, they form pairs and then migrate back toward the abdomen. Males copulate with females in the host body cavity. Males will remain there for the rest of their lives but fertilized female worms migrate forward again, through the lungs, and take up residence in the bird's air sacs in the upper chest. This allows them to get their eggs out of their host. Within approximately 170 to 180 days after infection, female tongue worms begin laying eggs, producing about 2,900 eggs per female.

It is a mystery why tongue worms go through this weird forward–backward–forward migration, but Banaja and colleagues hypothesized that the anterior air sacs have a much smaller volume then the abdominal body cavity, so the chances of two tongue worms encountering each other are better in the anterior air sacs. Essentially, the migration may be a mechanism to facilitate a *rendezvous* between male and female worms and to synchronize reproduction—the anterior air sacs are like singles bars for tongue worms.

The situation with tongue worms living in vultures may be different. In these birds, tongue worms occur in the abdominal air sacs and are immature. Spanish black vultures are non-migratory and feed on the carrion of small mammals (usually rabbits), but they also take live prey such as lizards. Perhaps the tongue worms of vultures use an intermediate host. Most tongue worms are also parasites in the lungs of reptiles, so vulture tongue worms may be using an alternate method of infection—from lizards to birds. More research is needed to clarify the life cycle of these tongue worms.

Every summer in Canada, after a long and cold winter, many people open their summer cottages and cabins. However, sometimes their joy at welcoming warmer weather is spoiled when their first nights in their cabin beds result in painful and irritating bites. A close inspection of the bedding usually results in the revolting discovery of what appear to be small, teardrop-shaped bed bugs. In fact, these bugs are most often not human bed bugs (*Cimex lectularius*), but rather, tend to be either bat bugs or bird bugs. Human bed bugs can attack birds that associate with us humans (like pigeons and house sparrows), and bird bugs can return the favour by biting people. At the lake, the bugs most often guilty of spoiling a good night's sleep belong to one of the 26 species of bugs in the family Cimicidae that live in the nests of birds, feeding off their blood.[6]

Bird bugs are insects belonging to the order Hemiptera. Hemipterans (often called "true" bugs) have mouthparts that are adapted for piercing tissue and sucking fluids. Most members of the order feed on plant sap, but these mouthparts also allow some bugs to feed on blood. Bird bugs are wingless, and they use their flat bodies to hide in the crevices of nests during the day. At night, the bugs become active and crawl onto a host. They use their mouthparts, which extend from the front of the head backward to the bottom of their thorax, to stab through the host's skin and suck its blood. Adult bugs can survive for months without a meal, patiently waiting in the nest for their migrating hosts to return. When people open their cabins after a long winter, these bugs have been waiting under the eaves in the nests of swallows or swifts, and they are hungry. They become activated and crawl along the rafters and drop onto the beds of their unsuspecting human victims below.

Bird bugs have been around for a long time. A specimen of a very primitive bird bug was found encased in amber in Myanmar (Burma), which dated back to the Cretaceous, 144 to 65 million years ago.[7] Because this specimen likely precedes the evolution of bats, it was probably a bird bug—or possibly a dinosaur bug? George and Roberta Poinar have suggested that during the Cretaceous, the proliferation of blood-feeding insects (and the diseases they transmitted) may have been an important factor leading to the demise of dinosaurs.[8]

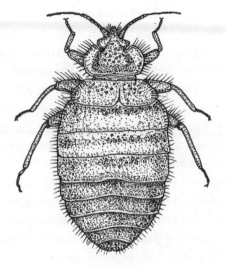

FIGURE 8.3 A swallow "bed bug." Cimicids are small, flattened bugs that hide during the day in crevices and feed nocturnally on blood. They attack birds, bats, and humans—species that share the habit of living in caves.

Only about 26 species of bird bugs occur—they have not diversified extensively. They are wingless insects and cannot easily disperse. To survive, they need hosts with permanent dwellings because both adults and all sub-adult stages need several blood meals to mature and reproduce. If their hosts are migratory, the bugs need them to faithfully return to the same nests year after year. Bats, which use the same roosts, and humans, who live in the same domiciles year after year, are suitable hosts.

Today, most blood-feeding bird bugs attack swifts (family Apodidae), swallows, and martins (family Hirundinidae)—birds that nest in colonies and often return to the same nests year after year. As a curious side note, humans are the only primates known to have a problem with bed bugs and it has been suggested that this stems from the time when we resided in caves along with swallows, martins, and bats.

One species of bird bug, the swallow bug (*Oeciacus vicarius*), infests the nests of cliff swallows (*Petrochelidon pyrrhonota*).[9] Cliff swallows are small migratory birds that nest in colonies throughout western North America. Cliff swallows construct gourd-shaped mud nests under cliff overhangs, the eaves of buildings, bridges, and in culverts. Colonies are usually located near water, which provides a ready supply of mud and emerging aquatic insects. Colonies can include anywhere from about 100 nests up to 1,300 nests. These social birds synchronize their nesting within a colony, and each pair usually produces one brood of young per year. After laying 3 to 5 eggs and incubating them for 12 to 16 days, the young are brooded for 5 to 10 days and are fledged within 22 to 25 days after laying.[10]

Swallows enjoy many advantages from nesting in dense colonies—protection from predators due to the vigilance of many eyes, better likelihood of finding food, and greater chances of breeding with genetically unrelated individuals, among others. One serious disadvantage of colonial nesting, however, is the large guild of blood-feeding ectoparasites that take advantage of a large and predictable source of hosts. In studies conducted in the central United States, cliff swallow colonies were found to be infested by four different species of blood feeders, including a short-lived, fast reproducing flea (*Ceratophyllus celsus*), two species of longer-lived, adversity-selected ticks (*Ixodes baergi* and *Ornithodoros concanensis*), and most commonly and abundantly, by swallow bugs.[11]

Swallow bugs can reach populations of more than 2,600 bugs per nest. Adult bugs lay their eggs on nest surfaces and in cracks in the mud, where the eggs hatch and undergo moults through five nymphal stages. All stages are blood feeders, so one nest may result in 13,000 bug bites (assuming that each bug takes only one bite per life cycle stage).

Like ticks (discussed in Chapter 4), these bugs are tough. They can remain alive in unoccupied nests for three to four years. When cliff swallows return from their migration to South America, hungry bugs congregate around the nest entrances, eager to feed. Occasionally, a swallow will inspect a nest but decide not to use it; however, this inspection provides the opportunity for a swallow bug to hitch a ride to the next nest.

The bugs have serious effects on swallows. In a study in which some nests in Texas were fumigated and compared to naturally infested nests, swallow bugs were found to shorten the fledging period by two days.[12] Moreover the fledged young from infested nests were 15% lighter in weight and had shorter wing and tail feathers than their uninfested peers. The mortality of young birds in parasitized nests was 76%, compared with 9% from the fumigated nests. In addition to weight loss, bugs can cause tissues to swell due to fluid accumulation and they may cause hyperemia, the pooling of blood. They also cause secondary bacterial infections and anemia, a decrease in red blood cells. And, if that weren't enough, swallow bugs also transmit Buggy Creek virus (a virus in the same group that causes western equine encephalitis).

Another study that compared fumigated and naturally infested cliff swallow nests found that fledged young from infested nests had twice as much asymmetry of wing and tail feathers.[13] Asymmetrical feathers pose a risk to swallow survival because they affect flight maneuverability and lift, which in turn alters foraging efficiency.

Research conducted over a six-year period in western Oklahoma confirmed the downside of communal living. It found that larger colonies of swallows had more swallow bugs per nest, and that the bugs tended to concentrate in the central area of a colony.[14] Bugs affected colony site selection, the birds' ability to synchronize nesting within the colony, and the likelihood that a nest (or a whole colony) would be abandoned. Measured against the benefits of nesting in colonies, cliff swallows often pay a heavy property tax to swallow bugs.

Swallows can try to mitigate this tax is by establishing new colony sites, but this is expensive due to the time and effort required. They can also try using old, formerly abandoned sites, but a few tough bug survivors may still lurk there, and new bugs will probably be introduced as hitchhikers anyway. Another tactic is to modify the breeding synchrony within the colony, so that early nesters pay a smaller tax than late nesters, or so that birds might have a second brood—a very costly and dangerous strategy because of the possibility that there will not be enough food to raise chicks. As a last resort, swallows may simply give up, abandon the colony (and their unfledged chicks), and gamble that next year will be better.

The last oddity in the flying zoo may not be that common on birds around the world, but it is a type of parasite that many humans are very familiar with. When I was young, I spent many hot summer days playing in a small creek near my parents' home in southern Ontario. My friends and I would wade in shallow areas looking for crawfish, frogs, and turtles, and we would float down the creek on inner tubes. Once, after coming out of the water, I noticed what I thought was a blob of mud on my leg. I casually wiped it with my towel, but it didn't come off. Then the blob moved—I was being attacked by a leech.

Leeches (class Clitellata; subclass Hirudinea) are annelid worms, related to earthworms. The first leeches were probably not parasites, but lived on the surfaces of crayfish or crabs in freshwater. Some may have become occasional ectoparasites, taking blood meals from fish. As they evolved to be more parasitic, leeches developed adhesive suckers that permitted them to attach to hosts and to feed for longer periods of time. Today, some are predators of small invertebrates, but of course, the notorious ones, as I discovered, are blood-feeding parasites. Many species of leeches are found feeding on the exposed skin of waterfowl, but three blood-feeding leeches (*Theromyzon rude*, *T. tessulatum*, and *T. biannulatum*) feed in the nasal passages, trachea, and beneath the nictitating membranes ("third" eyelids) of birds.

These leeches do not show much host specificity (many types of waterbirds can be attacked), but dabbling ducks and swans seem to be the most frequently infested. The leeches appear to be restricted to cold-water lakes, and they peak in the late spring and early summer months.

Birdwatchers and hunters may notice amber or olive-coloured leeches protruding, blood-engorged, from the eyes and nasal openings of waterfowl. Infested birds may scratch their heads, rub their heads against their bodies, shake their heads, and even sneeze to try to dislodge the leeches. Usually, these efforts are unsuccessful because as anyone soon realizes (and as I discovered when I used my towel), leeches have well-developed muscular suckers that allow them to stay firmly attached to their host.

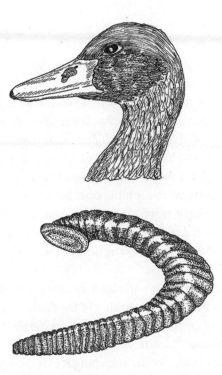

FIGURE 8.4　A leech (*Theromyzon rude*) in the nasal passages of waterfowl. Here, a mallard duck has a leech protruding from its naris (nostril). Leeches are highly adapted annelid worms that attach to their host using muscular suckers.

Leech infestations of waterfowl from northern lakes are common. Hungry leeches move toward light (unlike fed ones, which move away from light) and are attracted to heat, vibrations, and preen gland secretions.[15] Flocks of hungry leeches can sometimes be seen congregating near the surface of a lake. When leeches infest the eyes, they attach to the conjunctival membranes, beneath the nictitating membrane, where they cannot be scratched out. They seldom attach to the cornea and seem to have regulated populations (at least as adults, though not as juveniles), resulting in one leech per eye. This seems to cause little interference with the bird's vision. Nevertheless, leeches can cause the cornea to become opaque and the eye to become swollen. In the nasal passages and trachea, leeches attach to mucous membranes and engorge on blood. They can block air passages and cause inflammation and damage to respiratory tissue linings.

There has been at least one report of leeches being associated with the deaths of trumpeter swan cygnets (*Cygnus buccinator*) and other young waterbirds.[16] Usually though, it seems that leeches seldom cause significant disease to waterfowl, but if birds are sick due to another infection (such as botulism), then leeches can become abundant and cause greater discomfort to their hosts. Although the leech I had attached to my leg did not cause any pain or discomfort (leech saliva contains anesthetics and anticoagulants), I have to say that the thought of having my eyes, nasal passages, and throat infested is rather creepy. And leeches can carry other parasites, such as blood-dwelling trypanosome protozoans (some of which cause sleeping sickness in humans).

Not all leeches live in water—the land-dwelling varieties often make themselves known to researchers working in the field. For example, in 1903, the naturalists W. Dunlap and Leonard Cockayne visited Taumaka, the largest of the Open Bay Islands, which are located near the west coast of New Zealand's South Island.[17] While searching for wetas—grasshopper-like insects that are among the largest on Earth, at lengths of 10 cm and weights of 70 g—Dunlap reached into the burrow of a sooty shearwater (*Puffinus griseus*) and was bitten on the wrist by a leech. More searching revealed six more specimens and some cocoons. The leeches were identified as a new endemic species, *Hirudobdella antipodum*.

Terrestrial leeches from the rainforests of Southeast Asia and the Pacific have been known for some time. Many live on the wet ground but some also occur in the canopy of the forest. Some Australian species in the rainforests of Queensland feed on marsupials and birds and probably attack invertebrates as well. They can also transmit trypanosomes.[18] I once heard of a biologist working in Australia who had leeches dropping on him from the canopy above. After getting back to camp and removing his hiking boots, he discovered that the leeches from the forest floor and canopy had fed wherever the eyelets of his boots were, leaving a perfect pattern of bite marks on his feet and ankles.

Sometime around 1910, flightless endemic New Zealand birds called weka (*Gallirallus australis*) were introduced to the Open Bay Islands. Weka are chicken-sized omnivores that eat many kinds of invertebrates as well as seeds and fruit. By 1969, researchers could no longer find the leeches on the islands and it was presumed that weka had driven them to extinction.[19] Then, in 1987 a photographer crawled under a glacial boulder to take photographs of Fiordland crested penguins (*Eudytes pachyrhynchus*), and a leech attached to his eyebrow. Subsequent searchers discovered that a small population of the leeches still exists—so small that they are now considered an endangered species.

The Open Bay Island's leech must be a stubborn and resilient species. The supply of vertebrate blood for any terrestrial leech is normally scarce and sporadic—for these leeches, the food problem must be much worse. The crested penguins arrive on the islands in July to breed, and adults and fledglings leave in November.[20] They return in February to moult but leave again shortly thereafter. To compensate for the penguins' absence, the leeches probably take advantage of opportunistic feeding, taking blood from other birds, seals, and perhaps from resident frogs. In addition, they likely feed on other invertebrates like worms, though this food supply has been decimated by the weka.

Biodiversity is biodiversity, so species like leeches should be offered the same degree of protection as more glamorous species, like whooping cranes. New Zealand wildlife authorities have suggested that the population of weka be relocated away from the islands. The Maori owners of the Open Bay Islands, however, revere the weka and point out that the birds themselves are now considered a vulnerable species and must also be protected. This leads to the problematic situation in which an endangered species, the leech, (one that most people would consider to be disgusting) might be protected by removing another vulnerable species (the revered weka). The leeches do not have the public appeal of beautiful birds, but little is known about them, and it seems to me they deserve to be protected. One lesson that recurs when we study nature is that the evolution of ecosystems has led to complex, interconnected parts, and that the loss of one of the parts (even as unappealing as a leech) can have unexpected and adverse effects on the whole system.

"Weird" animal parasites may also deserve attention for practical reasons. For instance, leech saliva consists of a cocktail of enzymes. One, a protein anticoagulant called hirudin, has been shown to prevent blood from clotting and to break down clots that have already formed. Since being discovered in leeches, it is now commercially produced and used to treat people with cardiovascular problems.[21] Other secrets could be awaiting discovery. The oddities in our flying zoo deserve the same study as organisms elsewhere. They have unique and specialized evolutionary histories, and probably some fascinating stories to tell.

FLYING ZOO BEHAVIOUR

BEAUTY AND THE PARASITIC BEASTS

Above the emerald canopy of the lowland Amazon rainforest in eastern Ecuador, the silence is torn by the raucous screams of a mated pair of scarlet macaws as they fly above the treetops and land on a bare branch less than fifty metres away. The sun sparkles off their crimson breast feathers, which contrast gorgeously with their yellow-green and cobalt blue wings. The birds stretch their necks and use their large beaks and sharp-nailed toes to preen their long, symmetrical, red and blue tail feathers until, startled by something unseen in the forest, they erupt from their perch and noisily fly off. It's a spectacular sight—the kind that makes it hard to deny that birds are among the most beautiful and fascinating creatures on Earth. After all, who can look at the lavish colours and extravagant feather ornaments of peacocks, or pheasants, or parrots, and not be amazed by their showy displays? And who can listen to the intricate, melodious calls of warblers and not be enraptured? How did birds acquire their beauty? And how did their songs come about and what do they mean? Many of us assume that birds acquired their alluring appearance and behaviours as part of their efforts to attract a mate. But is it possible that parasites are part of the story as well—that they can, at least in part, be responsible for the astonishing displays and behaviours we see in birds?

Parasites would have little effect on the behaviours of birds if they established a purely benign and peaceful co-existence with their hosts, as they would not create much in the way of selection pressure. However, as we know, this relationship is often far from benign. The parasites' main goal is transmission to another host. This is facilitated by reproduction—the more parasite offspring there are, the better the odds of some of them getting to another host. Sadly, this high parasite reproductive rate often results in host pathology, especially when diseases are transmitted as well.[1] But even when a bird–parasite relationship appears to be causing no harm, a heavy cost can be imposed by the energy toll of keeping parasites in check, whether through the host's behaviours or activated immune response. These subtler costs, as well as the more obviously devastating ones, can exert pressure on bird evolution.

Benjamin Hart made the case that parasites and pathogens are just as important as predators and foraging in shaping the behaviours of animals.[2] He listed five types of behavioural responses that hosts may have evolved in response to the occupants of the flying zoo. Some of these are transparently related to coping with the pressures of parasites—for example, birds have evolved behaviours to avoid exposure to parasites, or to allow, but control, exposure. Other behavioural patterns, such as anorexia and depression, result from infection. Still others are less obviously related to parasitic pressures—for instance, behaviours that try to assist infected mates or kin, and, even more subtly, reproductive behaviours that try to provide offspring with the best genes for effective parasite control. In short, as we will see in this chapter, parasites affect almost all aspects of birds' lives, from nesting to feeding to their sex lives.

Birds use many behaviours to try to avoid parasites, some of which we've already seen in previous chapters. Ground-dwelling birds, like geese and grouse, may avoid feeding in areas that are contaminated with feces and may contain a concentration of nematode eggs. Colony-nesting birds, as we saw for penguins in Chapter 4, may compete for central nest sites that have fewer parasites. And, as discussed in Chapter 6, great horned owls may spend more time on the forest floor to avoid blackflies that transmit malaria, and Hawaiian honeycreepers alter their feeding patterns and even their sleeping postures to avoid malaria-transmitting mosquito bites.

An ingenious means of controlling ectoparasites is evident in marine fish, which use cleaning stations where invertebrates such as shrimp or other fish pick off their ectoparasites. Such outsourcing behaviour has never been reported for birds. However, birds themselves may provide this cleaning service for other animals—for example magpies (*Pica pica*) in western Canada eat winter ticks (*Dermacentor albipictus*) from the backs of bison, and Darwin's finches feed on ticks on marine iguanas in the Galapagos.

Birds use anting behaviours, lying prostrate on an anthill or dabbing ants onto their plumage, to chemically control ectoparasites. Surprisingly, few studies have provided strong evidence for the effectiveness of this behaviour. On the other hand, there is much support for the idea that birds fumigate their nests with plants such as fleabane, wild carrot, and goldenrod. In a study where fresh plant material was removed from starling (*Sturnus vulgaris*) nests, the populations of northern fowl mites (*Ornithonysus sylviarurn*) greatly increased.[3]

Perhaps the most common—and effective—behaviours for controlling parasites are scratching and preening, which were discussed in detail in Chapter 2. Experimental studies show especially strong evidence for the importance of preening. For example, Sandra Brown "debeaked" chickens and infested each of them with 50 *Menacanthus stramineus* feather lice.[4] After 33 days, these chickens had 1,600 lice per bird, whereas normal-beaked chickens in the control group had fewer than 60. Handicapped birds also preened themselves 5 to 10 times as much, trying—unsuccessfully—to control their lice.

In addition to self-preening, the practice of preening other birds, or allopreening, plays an important role. Penguin studies showed that both paired and unpaired penguins self-preened for about 10 minutes per hour, but unpaired birds became infested by 3 times as many ticks, presumably due to the lack of a solicitous, allopreening partner.[5]

ATTRACTING A MATE

If you're a bird, acquiring a partner may help you stay parasite-free, but you might first need to convince a prospective mate that you are not overrun with parasites to begin with. No one, not even a bird, needs a project as a mate. It's in this process of mate selection that we can see some of the most intriguing effects of parasites on bird evolution.

Like many Victorian naturalists, Charles Darwin watched, killed, and collected many birds. He also kept and selectively bred fancy pigeons— a hobby that provided supporting evidence for his theory of evolution by natural selection, which predicts that traits that enhance survival will spread throughout a species. But Darwin also noticed that birds have some traits that appear to hinder their survival. For example, he wondered why wild birds sometimes trade protective camouflage colours for gaudy displays that hamper their movement and make them more visible to predators. He suggested that a trait like a peacock's tail, which seems to run counter to the principles of natural selection, must offer some mating advantage that outweighs the risk to survival. Darwin coined the term "sexual selection" for the evolution of seemingly harmful characteristics, like showy feather displays in birds.

Today, the theory of sexual selection provides a general framework for more specific hypotheses about various aspects of the ecology, life histories, and behaviours of birds. In the 1980s, Bill Hamilton and Marlene Zuk proposed the intriguing idea that certain traits, including gaudy bird coloration and the size and symmetry of feathers, evolved by sexual selection because they provided prospective mates with information about the health status of the displayer.[6] These traits might even provide clues about a bird's heritable resistance to parasites. Based on their hypothesis, a showy male bird was telling prospective mates, "Hey, look at me—I'd be a great dad because I'm hale and hearty and not too infected with parasites, so I can protect you and help look after the kids. Also, our kids probably won't get badly infected with parasites, and our sons will be really sexy too!"

One of the most appealing features of the Hamilton-Zuk hypothesis is that it can be tested by observations that relate parasite loads to bright colours or flashy displays. It is also amenable to experimental manipulation. Such observations and experimental tests set out to answer the following question: are sexually selected traits in male birds simply a way of attracting female attention and creating a good impression, or are they *true* advertisements, revealing something important about a male's fitness?

If the Hamilton-Zuk hypothesis is true, then we would expect to find that the species of birds that are most susceptible to parasites should be those that have evolved the most elaborate ornaments and displays—that is, there should be a positive relationship between the prevalence of parasites (the proportion of hosts that are infected) and the degree of ornamentation. This is because the high risk of infestation makes it all the more important for females to identify males that can resist the parasites, providing pressure on males to evolve ways of advertising their good health. And, within a species, just the opposite trend should be seen—birds that are the most heavily parasitized should have the poorest displays. In other words, a showy display should provide honest information about resistance to parasites.

Several correlational studies have supported the Hamilton-Zuk hypothesis. For example, Andrew Read studied North American and European passerines, and found a positive relationship between the prevalence of blood parasites in a species and the brightness of its plumage, even when he controlled for phylogeny. (This means that by looking only at passerines, like waxwings and creepers, rather than comparing more evolutionarily distant birds such as waterfowl and owls, Read could analyze the prevalence of parasites and relate this to male brightness without unduly biasing his study.)[7] Similarly, in a study of 526 species of neo-tropical birds, Marlene Zuk found that male brightness was positively associated with the prevalence of blood parasites.[8] This trend was especially strong for 451 species of birds that are considered residents of the neo-tropics (such as scarlet macaws) and do not migrate. This makes sense—resident birds are likely to have co-evolved with their parasites more intimately than have migrants, and the evolution of their plumage is just one manifestation of this close relationship.

WHAT DO FEMALE BIRDS LOOK FOR?

Anders Moller conducted an especially detailed study that extended over four years. He chose to look at the ectoparasites that infest barn swallows (*Hirundo rustica*).[9] These birds make a good target population for evaluating the Hamilton-Zuk hypothesis. First of all, barn swallow males compete vigorously for mates. The tail feathers of male swallows are about 16% longer than those of females, and the males use their songs and tail feather displays to attract female mates. Even after pairing, males will attempt to copulate with other females, who may accept or reject these "extra-pair" copulations.

Second, barn swallows are semi-colonial nesters that are highly susceptible to parasites. Moller found that they were infested with tropical fowl mites (*Ornithonyssus bursa*), nasty blood feeders that feed around the birds' heads and lurk in their nests. These mites are known to have a significant effect on the fitness of swallows—they reduce breeding success, reduce fledgling weight and nestling survival, and cause delays in second clutch attempts. The swallows were also infested with the feather lice *Hirundoecus malleus* and *Myrsidea rustica*, which do not seem to directly reduce fitness but may transmit other parasites and diseases.

Moller found that levels of ectoparasites were predictable over time: birds that had high parasite loads in one year had high parasite loads in subsequent years. This suggests that some birds have a heritable resistance to ectoparasites, a valuable trait that would be worth advertising to choosy females. Indeed, females clearly did take into account their mates' degree of infestation—whereas 82% of unmated males had mites (averaging 11 mites per bird), only 32% of mated birds were infested (averaging 6 mites per bird). And there was more direct evidence that parasite load was a factor in females' mate choices, rather than lower parasite load in males being the result of having a partner to help preen them. Males that were already heavily infested before mating took longer to acquire a mate (and in fact, males also preferred to mate with females that were less heavily infested with ectoparasites).

What information were females relying on to determine the parasite loads of prospective mates? Barn swallows with the longest outermost tail feathers had the lowest number of mites and lice. And, as Moller noted, female birds preferred males with longer tail feathers for extra-pair copulations. Hence, males that succeeded at being the most promiscuous were the ones with the longest tails—and the ones that often lacked lice and mites. (The connection between the length of tail feathers and parasite load may be due to the fact that parasites negatively affect swallows when they are in their winter quarters, the time at which the birds moult and the lengths of the tail feathers are determined.)

Moller also conducted an experiment to test the idea that females considered tail length on its own in choosing their mates.[10] He captured barn swallows and randomly divided them into one of three groups: for one group, he shortened the tail feathers of males by cutting two centimetres; for the second group, he glued on the tail pieces he had cut to give these male birds an extra two centimetres of tail length; and for the third group, he cut off tail feathers and then reattached them, to determine if simply undergoing a cosmetic procedure caused an effect. Then Moller did breeding trials with the males and recorded the time it took before a female copulated with them.

On average, males with short tails had to wait 12 days before they got lucky, whereas males with normal tails took 8 days, and those with long, sexy tails copulated in a mere 3 days. These data seem to support the predictions of Hamilton and Zuk. They indicate that females use sexually selected traits as evidence about parasitic infections in males. However, the studies don't tell us why females choose less-infested males. There are three possible reasons for this: Perhaps females avoid parasitized males so that they themselves won't get infected. Or perhaps females expect that parasitized males are less likely to help care for chicks or maintain good territories because they are sick or spending valuable energy to ward off their parasites. Or perhaps, as suggested by Hamilton and Zuk, females avoid parasitized males because their chicks might inherit inferior genes for parasite resistance. Moller's study of barn swallows is consistent with any one of these explanations.

However, the breeding behaviours of other birds may shed more light on what drives females' choice of uninfected males. An interesting example can be found among sage grouse (*Centrocercus urophasianus*), which hold mating assemblies called leks, where males strut and display inflated esophageal air sacs to compete for females. Choosy females will mate with only a few lucky males, while many cocks will not succeed in mating at all.

Linda Johnson and Mark Boyce studied sage grouse at eleven leks located in southeastern Wyoming.[11] They found that these birds were commonly infected with a variety of blood-dwelling parasites (for example, *Plasmodium*, *Haemoproteus*, *Leucocytozoon*, trypanosomes, and larval stages of nematodes) as well as by feather lice (*Goniodes centrocerci* and *Lagopoecus gibsoni*). These infections could lead to serious disease—for instance, *Plasmodium pediocetii* causes malaria, in which the red blood cells of birds may burst, especially in the morning, flooding the circulatory system with toxins and cellular debris.

Unlike lice, malaria and the other blood parasites cannot be transmitted directly between birds—females cannot be infected simply by mating with an infected male. Hence, if female sage grouse reject males with these parasites, it is not because they are trying to avoid becoming infected themselves. Moreover, males play no role in helping to raise the chicks, investing nothing in parental care. Therefore, parental fitness is also not a factor in female rejection of infected males.

Johnson and Boyce found that males with malaria did provide females with cues about their infection status. Although they strutted as frequently as males without malaria, they attended the leks only about half as often. (Malaria parasites only break out of blood cells on some days, leaving the males well enough to attend leks on others). And females seemed to pick up on these cues; uninfected males copulated more often than expected, whereas infected birds copulated less often. Females seemed to be most impressed by males that attended the leks frequently, and they tended to ignore

FIGURE 9.1   A female sage grouse (*Centrocercus urophasianus*) at a lek judges two males
as prospective mates. One male has several red hematomas on its inflated, olive-coloured
esophageal air sacs, which are caused by feeding lice (*Goniodes centrocerci* and *Lagopoecus gibsoni*).

parasite-infected males, even though these males danced and strutted as
vigorously as uninfected birds when they did attend leks. Females have good
reason to avoid males with malaria, despite the fact that it poses no threat to
themselves and has no impact on the raising of their young—the fertility of
infected males is impaired, because malaria reduces sperm motility and via-
bility.[12] If females mate with parasitized males, not all their eggs may hatch.

Male sage grouse who succeeded at breeding were also much less
likely to have lice (12% of breeding males versus 42% of non-breeding
males). Males with lice often have hematomas, or blood clots, on their air
sacs where lice have fed. These hematomas were readily visible to females,

especially when air sacs were inflated and males were strutting. Females used the spots as clues to detect lousy males. To test this idea experimentally, Margo Spurrier and colleagues painted red splotches on the air sacs of captive male sage grouse to mimic hematomas, and found that these males were chosen by females 32% less often than males that had not been cosmetically altered.[13]

THE SONGS OF HEALTH

Male birds have methods other than colourful or showy displays for attracting females. Many birds use complex songs to communicate, and it appears that song is a sexually selected trait, as female songbirds prefer males that sing more complex songs.[14] Do parasites affect the ability of males to sing in a way that allows females to recognize infected suitors?

Male birds learn their songs in the first few months of life. Song complexity in most birds is controlled by the volume of nerve cell nuclei in the forebrain called the high vocal centre.[15] The energetic costs of developing and maintaining this area in the brain are high, and stress of any kind (such as a shortage of food during development) disrupts the normal growth of the nerve cells and affects the quality of a male's song and his ability to win a mate.

Karen Spencer and her colleagues showed that parasite infection can result in simpler songs by male canaries (*Serinus canaria*).[16] They infected 12 young male canaries with malaria (*Plasmodium relictum*) just at the time when the main nuclei of the song control pathway were growing. A control group of 8 birds received saline injections at the same time. As expected, malaria had a significant effect on the physiology of the canaries—it reduced their red blood cell levels by 17% compared with the control group. Infection also had a major effect on canary songs. There was a 36% reduction in the volume of the high vocal centre part of the brain, and a 12.5% decline in song complexity. This experiment shows that parasites can influence a sexually selected trait and supports the idea that complex songs in birds can be honest indicators of male health and parasite infection.

A study of sedge warblers (*Acrocephalus schoenobaenus*) in a field setting in England sheds further light on these ideas.[17] In this study, researchers looked in detail at the relationships between the song characteristics of
sedge warblers, their levels of parasitism, and the parental care given by
male birds.

Sedge warblers in this area were found to be infected by three types of
blood-dwelling protozoans—*Haemoproteus*, *Plasmodium*, and *Trypanosoma*,
which are transmitted by blackflies, mosquitoes, and louse flies.
Haemoproteus and *Plasmodium* live inside blood cells, ultimately destroying
them, which leads to anemia and anorexia in the host. In addition, the liver
and spleen of infected birds becomes enlarged and congested. Capillaries
become blocked and the back-pressure of blood results in fluid accumulation
in the lungs leading to pneumonia. Trypanosomes are flagellated protozoans
that can infiltrate birds when they eat infected louse flies (Hippoboscids).
Inside the bird, the trypanosomes live in the bloodstream and bone marrow.
Besides releasing toxic chemicals that affect the host's physiology, behaviour,
and endocrine system, the host's immune response sometimes causes a toxic
reaction that destroys its own tissues.

Near Surrey, England, 19.5% of breeding adult sedge warblers were found
to be infected by these parasites in 1995, and 37.5% were infected in 1996. Over
this two-year period, the presence of parasites reduced the song repertoires
of males by about 14% each year. In addition, in 1996, the researchers found
that on the day before birds paired, parasitized males spent less time "song
flighting," that is, engaging in energetic sexual displays of song, than did non-
parasitized males. Thus, the level of blood parasitism in sedge warblers was
related to the expression of two sexually selected song traits that are known
to be important in female choice: song repertoire and flighting behaviour.

The female sedge warblers were right to prefer uninfected males. The
parasitized males turned out to be poorer fathers and husbands, showing
less effort in provisioning nests. The authors of the study suggested that
parasitized males expend much metabolic energy on their immune systems,
draining energy away from other functions such as sexual displays and
caring for young. Hence, females that did choose an infected mate would
have to work harder to raise young.

Of course, the opposite cause–effect relationship should be considered: it is also possible that poor-quality males are more likely to become parasitized, in which case, the poor songs and poor parental investment of these deficient males would not be a direct consequence of their infections. However, examination of the immune response of infected birds (including an increase in the number of white blood cells) seemed to suggest that the poor condition of males was a direct cost of parasitism. It is still unclear though, if this occurs because infected males are directing so much energy to their parasite resistance that they have less available for parental care, or if parasitized males are simply too sick to help their mates.

The sedge warbler study lends further support to the Hamilton-Zuk hypothesis, which claims that sexually selected traits can signal good health and parasite resistance. However, an important component of this hypothesis is that resistance to parasites is also a heritable trait. A study of reed warblers (*Acrocephalus arundinaceus*) hints at the connection between genes, songs, and parasite resistance: it was discovered that the offspring of males with large song repertoires had higher survival rates than the offspring of males with smaller repertoires.[18] However, there does not seem to be direct evidence yet for the heritability of repertoire size, a link that is necessary in order to make a strong claim about co-evolution between parasitism and birdsong.

EXTERNAL SIGNS OF INTERNAL PARASITES

We've seen a number of studies that hint at a relationship between parasites and sexually selected traits. So far, all of these studies have involved either blood-borne protozoan parasites or ectoparasites like lice and mites. Can other kinds of parasites shape sexually selected traits? For example, is the presence of internal parasites, like worms, also signalled by traits that affect birds' choice of mates? If we find evidence that the Hamilton-Zuk hypothesis holds for a broad range of different parasites, this would suggest an even more important role for parasites in the evolution of sexually selected traits.

Red grouse (*Lagopus lagopus scoticus*) provide an interesting opportunity to explore this question. These birds live on the heather moor-
lands of Scotland. Unlike sage grouse, they do not use leks but are mainly
monogamous and defend territories from autumn to spring.

Red grouse can be plagued by infections of roundworms, or nematodes (*Trichostrongylus tenuis*), that live in the birds' caeca, the paired blind-
ended pouches that branch off from the large intestine. In high numbers,
caecal worms can cause inflammation and bleeding, reduce digestive
efficiency, and lead to host death. Heavily infected female grouse have
reduced fecundity. The birds are especially susceptible to this disease in
spring, when the nutrient quality of heather is poor, and when the birds
are stressed by breeding activity.

Adult worms lay eggs in the grouse's digestive tract. After the eggs
are passed out with bird feces and hatch, worms moult twice to the reach
the infective third larval stage and then migrate toward heather plants. If
eaten by a grouse at that point, they moult again to the adult stage and live
in the bird's caeca.

As the grouse population increases, so does the number of larvae on
the heather, reaching levels so high that every bird in a population can
become infected. In one study, the average number of caecal worms in red
grouse was 2,741, and it was not uncommon for older birds to be infected
with more than 40,000 worms.[19] (Other birds, like game-farmed pheasants, may use caecal worms, to which they are resistant, as a weapon
of competition to eliminate grouse from the same territory.)[20] Because
this parasite is not transmitted from bird to bird by direct contact, and
because male red grouse do not participate in raising young, the relationship between red grouse and caecal worms is a good place to look for
evidence of parasites affecting sexual selection; if females choose less-
parasitized males, it would be because these males have better resistance
to parasites, and not because the females are trying to avoid becoming infected themselves, or are concerned that the male may make a
bad father.

What external cues are there to tip birds off to the presence of caecal
worms in a mate?

Astute birds might look to the bright red combs that both male and female birds have over their eyes. Male birds with bigger and redder combs are more aggressive and more successful as breeders than males with smaller, less-conspicuous combs. However, the appearance of these sexual ornaments is determined by a number of different factors, so the information that they convey to potential mates is not at all straightforward.

A study done in northern England and Scotland showed that the combs of red grouse reflect two kinds of light—red light from the visible spectrum as well as light near the ultraviolet (uv) part of the spectrum (at a wavelength of 300 to 400 nanometres) that we cannot see, but that birds can.[21] Male grouse have bigger and redder combs than females, but their combs reflect less uv light. The size and brightness of male combs are greatest in the fall (when males establish their territories), and although young males have smaller combs, these are brighter than those of older males.

Visual information about birds' parasite loads seems to be related to one specific aspect of the combs. The researchers found that uv reflectance, but not total brightness or redness, was predictably related to the number of worms inside a bird—the higher the uv reflectance from the combs of males and females, the fewer the worms inside the bird. Red grouse are mostly monogamous, so uv reflectance could be an important source of information for both males and females, indicating how wormy their prospective mate is. In this system, where both males and females have a significant investment in choosing a good mate, such information could help each partner to increase its breeding success.

However, the story of red grouse, their combs, and their worms is a complicated one, and shows that information about a prospective mate's parasites is embedded within a network of desirable traits and their external signs. One complication is that comb size and redness in male birds is related to levels of testosterone, such that males with bigger combs are more aggressive, more socially dominant, and hold bigger territories. These dominant male red grouse also have better immune systems, and hence, might be better able to cope with caecal worm infections.[22]

Another complication is that worms also affect the territorial behaviour of red grouse, in addition to affecting the uv reflectance of their combs. In one experimental study, researchers treated some red grouse with anti-helminthic medication to reduce their worm infections and compared their territorial behaviour to that of an untreated control group.[23] Over a two-year study, treated birds won 92.5% of their territorial disputes, whereas untreated birds won only 67.5%. Perhaps worms reduce aggressiveness in grouse. This idea is supported by another experiment in which treated and untreated birds heard pre-recorded flight calls from an unfamiliar male. In response to this challenge, 80% of treated birds performed song flights, but only 25% of untreated birds did.

It appears, then, that a female red grouse get two different, but related, types of information from male combs—its size and redness tell her which male may be able to provide a good territory (and access to the resources it contains), while UV reflectance tells her how many debilitating parasites a male is infected with. Parasites also reduce male aggressiveness, and probably interfere with establishing and keeping high-quality territories, offering more direct evidence for a male's ability to provide. In turn, males also get information about the health of a female grouse and her ability to produce healthy chicks, information that is very useful for a monogamous male bird.

SEXUAL SELECTION FOR "GOOD" PARASITES

In all the cases we've seen so far, the parasites that affect the sexually selected traits of birds are pathogenic or at least cause their host to divert valuable energy away from the next generation. Their potential for harm is what motivates prospective mates to look for telltale signs of infection. But what if the parasites are not harmful or are even helpful? Although the claim is controversial, many scientists think (as discussed in Chapter 5) that this is true of feather mites living inside the quills or between the barbs of the feathers.

Like us, birds have a wide assortment of bacteria living on their surfaces. Some bacteria can secrete the digestive enzyme keratinase, causing the breakdown of feathers. Others may be harmless, or even beneficial, because of the competitive pressure they put on harmful bacteria—for example, they may take up space that would otherwise be colonized by feather-degrading bacteria, or they may release chemicals that can kill or slow the reproduction of harmful bacteria. Additionally, chemicals found in preen gland oil may selectively help or harm the bacterial flora, thereby affecting the appearance of the feathers. On the surface of a bird, there is a complex interplay going on between bacteria, mites, and hosts.

Feather mites may benefit their hosts by eating bacteria, fungi, and debris that is trapped in preen gland oil. And this could have a direct effect on the birds' appearance—by consuming old preen gland secretions with its trapped dirt and bacteria, and helping their hosts control feather-degrading germs, mites might actually improve the quality of their hosts' plumage. In this case, a bird might look for a mate with brighter plumage, not because bright plumage signals a *lack* of parasites, but because these birds likely have *more* of the helpful feather mites than birds with duller, dirtier feathers.

Ismael Galvan and Juan Sanz investigated this idea by studying great tits (*Parus major*) in Spain.[24] They discovered that the size of a bird's preen gland was an indication of the number of feather mites on the bird—the bigger the gland, the greater the number of mites. In addition, Galvan and Sanz measured the brightness of the birds' plumage and found that the bigger the preen gland, the brighter the bird. Finally, they noted that both large glands and bright plumage indicated an abundance of feather mites. Superficially, this appears to contradict the Hamilton-Zuk hypothesis that desirable traits signal good genes for health. However, if feather mites are mutualists and help birds have bright, clean, and healthy feathers, then this study supports the hypothesis at its core: the desirable trait is still conveying information about the bird's health, whether this involves the absence or presence of certain parasites.

The question of whether sexually selected traits can signal the presence of beneficial parasites is still an open one. Not all studies reach the same conclusions as the one by Galvan and Sanz.[25] And it is tricky to draw cause-and-effect conclusions from correlations between plumage brightness and numbers of feather mites. Certainly, complex relationships between birds, preen glands, bacteria, and feather mites are just starting to be unravelled.

More research in this area should shed light on the nature of the relationship between a bird's beauty and its parasites.

VISUAL SIGNS OF STRONG IMMUNITY

To evaluate the hypothesis that sexually selected traits advertise good genes for parasite resistance, one research strategy that we've seen is to look at the connection between certain desirable traits and the number of parasites on birds that exhibit these traits. But in addition to the challenge of figuring out if specific parasites are good, bad, or indifferent, other problems arise with this particular strategy.[26] Most kinds of parasites are not distributed evenly among all the members of a host population, but are usually over-dispersed—that is, a few birds are infested with a great many parasites, while most birds have few or no parasites of the same species. In this scenario, it's quite possible that a male bird with poor genetic parasite resistance has simply not been exposed. A female that relies on signs of infection might choose to mate with this bird, believing him to be a better mate than he really is. (And a researcher who does the same might be led to draw incorrect conclusions.)

Moreover, birds are seldom infected by just one species of parasite—normally the flying zoo is well occupied, and parasites vary greatly in their ability to cause disease. Hence, not all parasites have consequences that are relevant for evaluating a mate. There are also timing issues—perhaps a parasite infection during moulting will result in a poor display on a bird that is no longer infected during the mating season. One way to overcome these problems is to look for evidence of the host's disease resistance system, and especially the immune system, rather than looking for evidence of parasites.

How might a bird signal strong immunity? Ornamental plumage displays that include yellows, reds, and oranges usually involve pigments called carotenoids. Besides providing colour, some carotenoids are vitamin A precursors and play an important role in increasing the effectiveness of the immune system.[27] Pigments are also involved in "scavenging" dangerous chemicals, called oxygen radicals, that are unavoidably generated by the metabolic functions of cells. In this role, carotenoids are considered anti-oxidants. It is even possible that birds may store such pigments in their livers, so that the dangerous oxygen radicals they generate can be unleashed as a weapon against parasites.

Carotenoid pigments cannot be made by birds—they must get them from their diet. If foods containing carotenoids are scarce, birds with vivid colours may be signalling that they have access to these limited, important chemicals and that they have so much pigment that they can maintain a healthy immune system and control parasites, and still be able to spend valuable pigment on a flashy display. In this case, carotenoid ornaments are signalling immunocompetence, the ability of a host to ward off or control parasites, and therefore have fewer (or less virulent) parasites.[28]

Females can adjust the amounts of carotenoids they allot to their eggs, and consequently, how well protected from parasites a future chick will be.[29] Mother birds may even compensate for mating with an unattractive, poor-quality male by diverting more antioxidants into egg yolks. For example, Kristen Navara and her colleagues discovered that female house finches (*Carpodacus mexicanus*) that laid eggs after mating with males bearing drab red and yellow feathers put 2.5 times more vitamin E and other antioxidants into their eggs.[30] This makes perfect sense if Hamilton and Zuk are right, and the drab males are in fact genetically inferior for disease resistance. It also suggests that bird–parasite co-evolution may be involved in a trade-off game, where birds must balance investments in parasitic disease resistance against the pressures of reproduction.[31]

Such a trade-off effect was seen in a study of barn swallows by Nicola Saino and her colleagues.[32] The researchers injected male swallows with sheep cells to create an immunological challenge and altered the lengths of tails of some of the males. Birds with artificially long tails produced fewer antibodies than either control birds, whose tail length had not been changed, or swallows with short tails. In another study, researchers developed lines of domestic chickens that had low, high, or normal antibody production when challenged with sheep red blood cells.[33] Chickens that produced many antibodies when challenged in this way had smaller sexual ornaments (combs) than other birds, supporting the idea of a trade-off between immunity and display.

Besides carotenoids, other molecules play a role in colourful displays of birds. In parrots such as scarlet macaws (*Ara macao*), there are special types of pigments called psittacins or parrodienes.[34] The source of these chemicals is unknown, but they are responsible for the full spectrum of colours—the reds, pinks, yellows, oranges, and greens—that make parrots the most colourful group of birds in the world. Research on burrowing parrots (*Cyanoliseus patagonus*) shows that the ornamental colours convey information about a bird's quality.[35] Red psittacins can act as antioxidants and are probably important immune system boosters. However, there is no research yet looking at correlations between the colours, pigments, immunocompetence, and parasites of parrots.

Melanin pigments, unlike carotenoids, are made by birds and deposited in the skin and feathers. Production and deposition of melanin depends on a bird's body condition during the moult.[36] In a study of great tits (*Parus major*) in Switzerland, Patrick Fitze and Heinz Richner infested some birds with forty blood-feeding hen fleas (*Ceratophyllus gallinae*), and found that the size of a melanin-based breast stripe was smaller in infested birds compared with uninfested birds.[37] The width of the breast stripe (though not the blackness of the stripe or the coloration of carotenoid-based plumage) proved to be an honest advertisement of previous exposure to parasites. Melanins are also thought to be involved in immune system function—their production produces oxygen radicals that are scavenged by carotenoids.[38]

Different pigments may be signalling somewhat different information about a bird's parasites. Overall, it seems that carotenoids are good indicators of general bird health and condition, and signal infections by endoparasites, whereas melanins provide information about ectoparasites.

COMPLEX CHOICES

Over evolutionary time, the association between parasites and sexual signalling becomes more and more fine-tuned. The best example of this can be found among satin bowerbirds (*Ptilonorhynchus violaceus*) in Queensland, Australia. These birds have an unusual mating behaviour in which males

FIGURE 9.2 A male satin bowerbird (*Ptilonorhynchus violaceus*) builds and decorates a bower to attract a female mate. The size and symmetry of the bower, and the number and flashiness of decorations (feathers, flowers, and snail shells), are related to the size of the male and to the number of lice (*Myrsidea ptilonorhynchi*) with which he is infested.

build elaborate bowers, which are ʋ-shaped structures made of twigs. They decorate the bowers with colourful flowers, berries, feathers, and even human-made objects. Some males steal particularly nice ornaments from other bird's bowers. Stephanie Doucet and Robert Montgomerie discovered that the quality of a male's bower—its symmetry, the size and density of the twigs used, and the number of decorations—was directly related to the size of the male bird that made and decorated it.[39] Larger males built better bowers and were more successful thieves. Even more surprising, however, was the fact that the quality of a bower was related to the number of blood-feeding lice (*Myrsidea ptilonorhynchi*) found on the head of its builder. The better the quality of a satin bowerbird's bower, the fewer lice it had.

In addition, bowerbirds have iridescent plumage and reflect UV radiation, which can be detected by females. Doucet and Montgomerie found that the brightness of the UV rump plumage of males was directly related to a male's body size and the degree to which he was infected with *Haemoproteus*, a parasite that causes malaria. Again, the brighter the plumage, the fewer the parasites.

Female satin bowerbirds sequentially eliminate males as prospective mates—first, they check out the bowers and eliminate males who have built shabby structures, and then they inspect the birds themselves and choose the males with the brightest plumage. Females assess at least three different signals, getting information about male body size and levels of infection by two different kinds of parasites. Female satin bowerbirds have larger brains than males—possibly as a consequence of having to make complex evaluations involving bowers, male plumage, vocalizations, and displays. In this case, parasites may be affecting sexual selection and even braininess of birds.

After learning about the wide-ranging effects that parasites seem to have on bird displays, I can never look at or listen to a bird in quite the same way. Female birds obviously gain many benefits by pairing with males that have the best displays, because the flashy colours and beautiful songs are honest reports of a male's condition.[40] Overall, a female will never go wrong by choosing a male in good condition, and the reason why the male is in good condition (whether it is because he has good genes, a good immune system, few parasites, a great territory with lots of resources, or simply because he has been very lucky) really does not matter to her. What is important is that the signs of male fitness can change accurately and quickly depending on circumstances. For example, if a new, virulent parasite makes its way into a population, birds with the best resistance should advertise their superiority soon after by the expression of their ornamental traits.

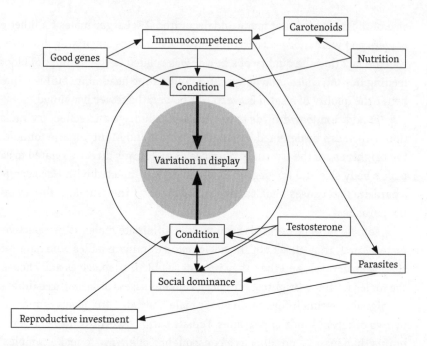

FIGURE 9.3 Some of the factors affecting the variation in displays of birds (songs, plumage colours, and behaviours), highlighting the possible roles of parasites. Note that not all possible connections are shown—for example, social dominance leading to the defense of a good territory could easily affect nutrition.

BIRD-ON-BIRD PARASITISM

Undoubtedly, the parasites that have resulted in the most significant and fantastic bird behaviours are the worst parasites of birds—other species of birds. More than a hundred different species of birds are brood parasites. Brood parasites sneak into the nest of a host (frequently a passerine), lay an egg, and leave it to be raised by the unsuspecting foster parents. Parasitic bird eggs tend to mimic host bird eggs in shape and colour, but have thicker shells and shorter incubation times. After hatching, the parasite nestling (for example, a common cuckoo, *Cuculus canorus*), will often eject the host's eggs from the nest. Honeyguides (*Indicator indicator*) that have infiltrated a host's nest use their hooked beaks—while still blind and featherless—to stab host chicks to death. Even if the parasite nestlings let their nestmates survive, they are larger and can use the host's begging calls to entreat their foster parents to feed them, which often leads to the starvation of the host chicks.

Brood parasites are nasty parasites that affect two generations—adults and chicks. They can go to dangerous lengths to foist their eggs on foster parents. Marie Abou Chakra and her colleagues reported that brown-headed cowbirds (*Molthrus ater*) act like gangsters.[41] If a host ejects an intruder egg from its nest, cowbirds will take revenge by destroying the entire nest. Thus, they extort their hosts, making them an offer they can't refuse—namely, accept the parasite or else. Brood parasites drastically affect the Darwinian fitness of their hosts.

The behavioural adaptations that parasitic birds have evolved defy belief. Some scare their prospective hosts away from nests by disguising themselves as predators, so that they can gain access to the nest to quickly lay an egg.[42] Polymorphic Old World cuckoos (*Cuculus canorus*) have barred underparts and de-curved beaks that resemble Eurasian sparrowhawks (*Accipiter nisus*), an example of Batesian mimicry in which an animal imitates the appearance of a dangerous species. Using a very different strategy, female cuckoo finches, *Anomalospiza imberbis*, use a wolf-in-sheep's-clothing tactic. They disguise themselves as harmless birds (southern red bishops, *Euplectes orix*) to deceive their hosts, tawny-flanked prinias, *Prinia subflava*.[43] This strategy, however, may be failing because prinias now mob both disguised parasites and innocent red bishops.

In response, hosts have co-evolved many defensive behaviours to counteract brood parasites. Weaver birds (family Plocidae) such as the southern red bishop construct their nests with narrow entrances that make it harder for parasites to enter. There have been reports of parasitic cuckoos trapped inside such nests.

Some hosts have become meticulous in distinguishing their eggs from parasite eggs. In two families of African birds, wagtails (Ploceidae) and cisticolas (Cisticolidae), individual host females lay eggs with unique, unpredictable signature markings.[44] These birds can detect cuckoo or cuckoo-finch eggs by their lack of signature and push them out.

Other birds learn to recognize parasite nestlings and toss them from their nests or starve them.[45] Most amazing are Australian superb fairy-wrens (*Malurus cyaneus*); these birds teach their nestlings a secret password while the young are still embryos inside their eggs.[46] Fairy-wren chicks have begging calls that include parts of their mother's special incubation call. When parasitic bronze-cuckoo (*Chrysococcyx basalis*) nestlings do not use this password, they do not get fed. This occurs even if the bronze-cuckoo nestlings have markings that make them look identical to their hosts.

Because prevention is better than a cure, some birds have developed a neighbourhood watch program to warn others of brood parasites in the vicinity. In experiments with reed warblers (*Acrocephalus scirpaceus*) in England, researchers presented warblers with a female cuckoo model (or a control parrot model) while playing warbler alarm calls and painting a warbler egg to look like a cuckoo egg.[47] They found that warblers relied on the neighbourhood watch program (the alarm calls) as well as their own observations in the nest (the presence of the cuckoo model and the egg's resemblance to a cuckoo egg) in deciding whether to eject these eggs. This helped birds to avoid mistakenly ejecting their own eggs.

Successful brood parasites need to be smart, as do the hosts that try to thwart them. This may have led to an arms race in braininess. Among the parasitic brown-headed cowbirds, for example, the egg-laying females need to have good memories; they lay one egg in a host nest at dawn, then scout for more host nests to use a day or two later. They must remember the locations of host nests for up to eight weeks until they've laid forty or more eggs. David Sherry and colleagues compared the brains of males and females of this species and found that female brown-headed cowbirds had larger hippocampus areas, which are associated with spatial memory.[48]

Even these most dangerous parasites, however, can play the role of friend rather than foe, if the circumstances are right. For example, as we saw in Chapter 6, the giant cowbird (*Molothrus oryzivorus*) in the Amazon may help oropendolas and caciques reduce the impact of botfly parasites. As a result, the parasitic nestlings are tolerated. Similarly, Daniella Canestrari found that carrion crows (*Corvus corone*) in Spain showed no attempt to avoid being parasitized by great spotted cuckoos (*Clamator glandarius*), and in fact, the presence of cuckoo chicks in their nests enhanced the crows' breeding success by 40%.[49] Why? The parasite chicks release what Canestrari described as a "pungent, noxious, disgusting" fluid from their cloacas. This substance evidently looks like rotting matter and causes a burning sensation in the throat. When cats were offered meat that had been smeared with this secretion, they rejected it. In years of high predation pressure, these parasites can become mutualists.

Parasitologists have sometimes been accused of overstating the effects of disease organisms on all aspects of the lives of birds. On the other side, for a long time, some ornithologists, ecologists, and naturalists have tried to believe that parasites are unimportant, disgusting, sporadic, and uncommon creatures that, at best, play a minor and random role in the lives of a few birds in a small area. In this chapter, I have tried to show that the truth lies somewhere in between these two extremes—parasites *may* be influencing anything from the brilliance and complexity of bird displays, their reproductive success, and even their intelligence and many of their behaviours, but they are only a part of the overall picture.

Parasites can and do affect the evolution of their hosts through processes such as sexual selection. I think it would be a mistake for anyone interested in any aspect of the biology of birds to ignore the fact that parasites are a real part of the life of birds, whether or not we (or birds) like it. By considering the *possible* influence of parasites, just as Bill Hamilton and Marlene Zuk did, biologists will find a new world of intriguing questions and puzzles. These can only make the intricate song of a warbler, or the weird decorations of a satin bowerbird's bower, or the spectacular colours of a scarlet macaw, all the more beautiful.

10

ENVIRONMENTAL

IMPACTS

The Future of the Flying Zoo

NATURAL AND UNNATURAL CHANGES

In the words of folk singer and Nobel laureate Bob Dylan, "the times they are a-changin'." One of the great pleasures in the study of natural history is the observation of changes—for instance, the seasonal transitions from birth and regeneration in spring to death and dormancy in autumn, or the year-to-year changes in temperatures and patterns of rainfall, which have such powerful ramifications for life. Naturalists also note longer-term changes that unfold over many years, such as the replacement of a lake by marsh-land and eventually by a field or forest in a process called succession, or the colonization of a patch of tropical rainforest after one of the giant, buttress-rooted trees has fallen.

My own family loves to listen for the first morning each spring when we are rudely awakened by the frantic honking of Canada geese (*Anser canadensis*) flying low over our house to land on a lake nearby. To us, this is a sure sign of the change from cold, white winter to warm, green spring. These changes are part of the rhythm of nature. To me, there is something reassuring about natural changes—they tell me that although individual life forms change and die, organisms adapt and persist—life persists.

Parasites are a large part of this persistent life. They account for somewhere between 50% to 70% of all living things.[1] There may be more than 300,000 parasitic helminths alone, occupying just 45,000 species of vertebrates.[2] Parasitism is the most common way that organisms consume carbon and get energy. Parasites are crucial and ecologically important— for instance, about 40% of all species in salt marshes along the coast of California are parasites. When researchers carried out detailed analyses in this ecosystem of the feeding interactions between levels in the food web, they concluded that parasites "appear as hidden 'dark matter' that holds the structure of the web together."[3] What changes affect parasites, particularly those that use birds as hosts?

Like other ecosystems, the flying zoo undergoes seasonal and annual changes, as well as longer-term changes. For example, when red-necked grebes (*Podiceps grisegena*) return to freshwater prairie lakes in spring after spending the winter living and feeding in coastal marine areas, they have a very different community of helminth worms living in their digestive systems than they did before they left. It would be interesting to find out what niches and ecological roles these marine parasites fit into, compared with the worms that live in grebes at the end of summer, when the birds have been living, feeding, and reproducing on a lake in the prairies.

Broader, long-term questions also come to mind—for instance, do migratory birds, like prairie grebes that encounter many types of parasites, have better immune systems than flightless, non-migratory grebes that spend their entire lives on one lake high in the Andes Mountains? And how does the flying zoo change over a bird's lifespan, as it ages from a downy chick to a full-fledged adult with a brood of its own? These are natural changes to which birds and their parasites have adapted for millions of years.

Unfortunately, however, increasing evidence suggests that many changes now taking place in the flying zoo are not natural, but are consequences of the activities of our species, *Homo sapiens* (*homo*—"man"; *sapiens*—"wise"). The human population on Earth currently stands at over 7.5 billion and is growing exponentially, like money in a compound-interest bank account, by about 1.1% annually. This means that each year, there are about 83 million more people on our planet than there were the year before— and about an extra 450 people in the time it has taken you to read this chapter so far. During my own lifetime, our population has almost tripled.

Consequently, human activities and their effects on the natural world are increasing as never before. Exploitation and greedy use of resources pollute the water and air, destroy natural habitats, and have altered the global climate. While this book was being written, Seattle, Washington (normally one of the wettest cities in North America) broke a record by having 52 consecutive days without rain. During this drought, 15 days had temperatures of above 80°F (27°C), and air quality in Seattle was dangerously poor due to smoke from the 150 forest fires (also a record-breaking number) in Washington's northern neighbour, British Columbia. Climate changes are real, are happening faster than at any other time in history, and result from our actions.

We are currently living in the midst of one of the greatest episodes of species extinctions ever recorded—about 55,000 species of living things are lost each year, and two species have gone extinct in the time that it has taken you to read this chapter.[4] Not surprisingly, these effects are reflected by changes to the flying zoo. Unlike natural changes, which often develop slowly and predictably enough to trigger adaptive changes in organisms, these newer changes may cause rapid instability. They are disturbing and sad to someone interested in birds and their natural history, but more importantly, some will boomerang and have perilous effects on us humans.

ENVIRONMENT AND DISEASE

In the summer of 1999, five people were admitted to Flushing Medical Center in Queens, New York, complaining of fever, general weakness, and confusion.[5] Medical detectives discovered a pattern: these sick people all liked to spend lots of time outdoors on warm summer evenings—and all had been exposed to mosquitoes. In addition, large numbers of crows (*Corvus brachyrhynchos*) were being found dead in Queens as early as June of that year, and by September, a bald eagle (*Haliaeetus leucocephalus*), a snowy owl (*Nyctea scandiaca*), flamingos (*Phoenicopterus*), and cormorants (*Phalacrocorax*) had died at the Bronx Zoo. Samples sent to the US Center for Disease Control and to the US Army Medical Research Institute of Infectious Diseases determined that the birds had died from a flavivirus (a family of viruses that cause inflammation of the brain—encephalitis—including yellow fever and dengue fever). The virus in New York was West Nile virus (WNV). By the beginning of October 1999, it seemed clear that both humans and birds in New York had contracted WNV from mosquito bites.

WNV was discovered in 1939 in Africa and, until its outbreak in New York, was only known to occur in Africa, the Middle East, and southeastern Europe. The virus strain isolated from the New York cases shared 99.8% its DNA with a strain that killed a goose (*Anser*) in Israel in 1998. In addition to humans, the virus infects more than 200 species of birds, as well as horses and other animals.[6]

The vectors, or agents of transmission, of WNV are mosquitoes. The main types of mosquitoes that transmit WNV to birds are several species in the genus *Culex*. Unfortunately, at least 11 of these will take blood from other animals (including humans) as well as from birds. Thus, infection can be spread from birds to us, in a process called zoonosis, as likely happened in New York.

By 2004, the infection had spread from New York across most of North America and killed more than 450 people and countless birds, as well as horses and other animals. WNV has moved south and is now in Latin America and the Caribbean—it was found in birds in South America in 2005. It appeared in northern Argentina in 2006. Public health officials in New York originally hoped that winter conditions would eradicate the WNV mosquito vectors, but culicid mosquitoes survived the winter in sewers and abandoned buildings.

Because the virus needs biting arthropods for transmission (usually mosquitoes, but ticks or mites may also be vectors), epidemics of WNV in temperate areas of North America occur only in the summer and early fall. In more tropical climates, the disease occurs year-round.

Although humans and horses suffer from WNV disease, they are poor hosts for the virus because not enough virus particles accumulate to readily infect mosquitoes. On the other hand, many species of birds accumulate high concentrations of the virus and are infective for a long time. Common backyard birds, like blue jays (*Cyanocitta cristata*), robins (*Turdus migratorius*), and house sparrows (*Passer domesticus*) are good hosts for WNV, and can act as reservoirs of the disease—that is, they maintain the disease within an ecosystem. In addition to acquiring the virus through mosquito bites, birds also become infected by eating infected arthropods. Hence, the spread of WNV may be an important development among insectivorous birds like swallows and martins. Experiments with captive birds indicate that transmission can also occur directly by fecal–oral contamination, and by allopreening, when one bird eats the ectoparasites of another.

Outbreaks of WNV tend to occur in temperate areas after mild winters that are followed by dry springs and summers. These conditions help mosquitoes survive. Droughts kill mosquito predators (such as dragonflies) and concentrate pools of stagnant water, which are ideal mosquito breeding sites.[7] These conditions were in place in New York in 1999 and in Colorado in 2003.

Severe storms with heavy rains at the end of the summer can also amplify transmission of WNV. If current, human-caused global changes result in these types of climate events, then we are likely to see greater epidemics of mosquito-vectored infections like WNV. The consequences for many bird populations—including endangered species like whooping cranes (*Grus americana*)—are grim, and of course, the chances for human infections will increase as a result. A vaccine may be developed to protect us, but vaccinating large numbers of birds is not practical—and maybe not even desirable, because letting nature take its course could result in selection for resistant birds and may also lead to the evolution of less virulent strains of the virus.

Of course, if this scenario for WNV is accurate, it will be true for other vector-borne diseases. Bird malaria and other blood-dwelling parasites that are transmitted by biting arthropods are likely to become more widespread and will be introduced to birds that have had little previous exposure. These birds will have no innate resistance, being immunologically naïve, and likely, their populations will be reduced. Sadly, this unnatural experiment has already taken place in the Hawaiian Islands (as discussed in Chapter 6), where mosquitoes and bird malaria were introduced less than a hundred years ago.[8]

EVOLVING SOLUTIONS

Because they had no previous experience with avian malaria when it was introduced, native Hawaiian birds were very susceptible to infection; 65% to 90% of birds that became infected died, even if the infection resulted from a single mosquito bite. Populations of native birds such as honeycreepers were decimated, and native forest birds that lived below an elevation of 900 metres, where mosquitoes were the most common, were nearly driven extinct.

Follow-up research, however, led to a surprise. When Bethany Woodworth and colleagues surveyed the population of Hawaii amakihi (*Hemignathus virens*)—native honeycreepers that are very susceptible to malaria—in forests on Mauna Loa volcano below 270 metres of elevation, they discovered that these birds were more abundant there (making up more than 50% of the birds in one community) than in higher, disease-free forests.[9] Not only were amakihi located there, they were breeding there. Even more surprising, up to 40% of amakihi had malaria parasites in their blood, and more than 80% had anti-malaria antibodies, indicating that they were, or had been, infected.

These rates of infection prevalence are some of the highest ever reported—typically, in the Nearctic or Neotropics, somewhere between 2% and 4% of birds are infected. In Hawaii, not only were amakihi surviving with malaria, they were increasing their distribution and abundance in low-elevation forests, where disease transmission was greatest. How could they do this?

Woodworth and her colleagues found that mosquito vectors were common in low-elevation forests, and 15% of them carried malaria, so transmission to birds could occur all year long. The low-elevation forests of Mauna Loa may represent an ecological disease trap, where the quality of habitat and the abundance of food for honeycreepers is great, but where there is also disease and death. Could it be that lowland amakihi had evolved resistance to *Plasmodium relictum*, permitting them to repopulate the lowlands? What about birds from mid-elevation forests, where mosquitoes were less common and malaria transmission was more episodic?

The only way to find out if the high transmission rate of malaria in low-elevation forests had selected for disease resistance was to do controlled infections with birds not previously exposed to malaria from both types of forests. Carter Atkinson and colleagues experimentally infected low- and high-elevation amakihi with *P. relictum*, and found that the lower elevation birds suffered less weight loss, showed no decline in food consumption, and had lower mortality rates.[10] They concluded that the birds had developed tolerance (rather than resistance) to malaria. This is a rare documented case of co-evolution between a parasite and a bird host over a short time frame—in this case, about one hundred years.

The results of this study are exciting because they may provide clues for conserving native birds when they become exposed to novel parasites and diseases. Perhaps protecting habitats where small, relict populations of hosts still exist will be important to the long-term survival of rare endemic birds, rather than just focusing our attention on areas where a disease has not yet invaded. What does seem inevitable, however, is that birds and humans are going to be exposed to parasites with which they are unfamiliar—the inevitable result of climate change, animal range expansions, migration route changes, and accidental or intentional introductions by humans. One hundred years is a short time in evolutionary terms—can animals adapt fast enough to keep pace with the unmitigated rate of climate change that is now occurring?

ECOSYSTEMS UNDER STRESS

Nowhere on Earth are the effects of global warming more apparent than in the Arctic. The Arctic ecosystem is populated by bacteria, protozoans, algae, plants, fungi, and animals that have evolved adaptations to compete, survive, and thrive during periods of intense cold, when the sun may not be seen for weeks. This ecosystem is fragile, with small perturbations having the potential to cause large effects, because the specialized organisms that live in the Arctic are so highly adapted to this special environment. Changes or disruptions can result in many extinctions or replacements.

The Arctic contains little in the way of biodiversity, and the variety of different organisms gets smaller as latitude increases.[11] At the highest latitudes, biodiversity is lowest. High latitudes are dominated by a few specialized host species that tend to have fewer parasites and diseases than those nearer the equator. Global climate change seems to be causing the melting of the Arctic. Each decade, the region is warming by more than 0.5°C, sea ice is declining by more than 3.5%, and perennial ice is declining by about 11.5%.[12]

A review of 143 studies of different types of organisms, encompassing close to 1,500 species, indicated that about 80% had shifted their ranges northward.[13] It seems probable that Arctic birds are going to come under much more physiological stress due to their poor competitive ability, changes to migration routes, reduced ranges, reduced food sources, and mixing with unusual species they have never encountered before. Increased stress combined with immunological inexperience will likely create ideal conditions for Arctic residents to become infected by more kinds of parasites and diseases than ever before. Some of these trends are evident already.

For example, Frederick Wrona and his colleagues found that waterfowl have emigrated northward and now migrate earlier in spring and later in fall.[14] Shifts in the abundance of phytoplankton have resulted in misalignments between the birds' breeding cycles and the resources available to them, and there has been competitive exclusion of some northern species by southern invaders. It is even suspected that hosts use parasites they are tolerant to as weapons of competition.[15] Moreover, as the Arctic continues to warm, humans will likely introduce more domestic animals to the area, and their infections will likely spill over to the residents.

This problem of spillover from non-native to native species is already a concern in an ecosystem that is as far removed from the Arctic as you can get, the Galapagos Islands on the equator. The Galapagos Islands are famous as a natural evolutionary laboratory. These geologically recent volcanic islands are located in the Pacific Ocean about a thousand kilometres west of Ecuador. Charles Darwin and subsequent naturalists discovered that the islands are populated by many rare and endemic species—sometimes restricted to just one island—of plants, arthropods, and vertebrates.

The most famous and photographed of these are reptiles, including marine and land iguanas and giant Galapagos tortoises, but there are also 22 unique endemic species of land birds, including 13 species of Darwin's finches, 4 species of mockingbirds, a flycatcher, a dove, a rail, a hawk, and an owl.[16] Endemic waterbirds include a flightless cormorant (*Phalacrocorax harrisi*) and the smallest species of penguin in the world, (*Spheniscus mendiculus*). In addition, Española Island is the world's main breeding site for waved albatross (*Diomedea [Phoebastria] irrorata*)—about 12,000 pairs nest there, with a few other pairs known to nest on Isla de la Plata, Ecuador.

Almost all wildlife on Galapagos is extremely tame—you must be careful not to step on animals as you walk along the marked trails. Once, while trying to take a photo, I got pecked on the arm by a Galapagos penguin that was trying to hop into our boat. The uniqueness and scientific significance of the Galapagos Islands was formally recognized when they were designated as a UNESCO World Heritage Site.

Because of their historical notoriety from Darwin's visit, their unique scenic beauty, and their accessible and photogenic wildlife, thousands of tourists come to the Galapagos Islands; in 2018, more than 275,000 visitors came. The tourism industry is threatening to destroy the features that make Galapagos so wonderful. Besides causing noise, traffic, and pollution, tourists intentionally and unintentionally transport plants and animals that aggressively invade the islands and displace endemics. Tropical fire ants (*Solenopsis geminata*), blackflies (*Simulium bipunctatum*), guava (*Psidium guajava*), and rats are just some examples.

Seeds of introduced blackberries (*Rubus niveus*) are scattered by Darwin's finches, who love to eat them, and now threaten to extirpate an endemic plant, *Miconia robinsoniana*. Goats introduced by sailors as food have eaten giant tortoises out of their habitat. Park authorities and conservationists had to step in and implement goat control and captive tortoise breeding programs to save some species. Increasing numbers of visitors have stimulated the local economy, so more and more people want to live in Galapagos (in 2012, the resident population was 35,000). Residents and tourists need food, but shipping it from mainland Ecuador is expensive, so local agriculture, including fruits, vegetables, and domesticated animals, like cattle and chickens, has been permitted. This opens the door even wider to the parasite spillover problem.

Domestic chickens are raised on 5 islands in Galapagos (Santa Cruz, San Cristobal, Isabela, Floreana, and Baltra), and feral chickens can now be found on 4 islands (Santa Cruz, San Cristobal, Isabela, and Floreana).[17] A survey of these birds, both domestic and feral, showed that they were infected by 18 types of virus, 9 species of worms, 3 species of protozoans, as well as by lice and mites. Some parasites are specific to chickens, but alarmingly many can be transmitted from chickens to native birds.

Two potentially serious threats have been identified. Newcastle disease, caused by a virus, results in great mortality to both domestic and wild birds elsewhere. Gulls, cormorants, pelicans, and penguins seem to be types of wild birds that are very susceptible and, when infected, frequently die, having mortality rates of more than 90%.[18] Birds can get infected by inhaling virus particles or by ingesting water or food that has been contaminated by nasal secretions or feces of infected birds. On Galapagos, lava gulls (*Larus fuliginosus*), flightless cormorants, and Galapagos penguins all occur in small populations with little genetic diversity, so an outbreak of Newcastle disease among any of these birds would be devastating.

Roundworms (nematodes) belonging to the genus *Dispharynx* have also been found in chickens. This parasite has a broad host range and relies on intermediate hosts known as sowbugs (also called woodlice or pill bugs, or, scientifically, isopods), which are common in Galapagos. There is a relationship between rainfall and infectivity of stomach worms—more infections occur in dry environments, which favour isopods. The nematodes live in the proventriculus, which is part of a bird's stomach, and can cause severe disease, especially in young birds. They attach to the wall of the proventriculus, causing ulcers. The stomach then fills with white mucus and sloughed stomach tissue, which can block the passage of food and cause birds to lose weight and become emaciated. Again, gulls, cormorants, pelicans, and penguins are species that are at great risk.

So far, none of the protozoan blood parasites that cause avian malaria have been found in either wild or domestic birds in the Galapagos Islands, but the usual mosquito vector, *Culex quinquefasciatus* (also a vector of West Nile virus), was introduced in 1985.[19] Transmission of bird malaria may not be a threat now, because the normal desert conditions in the islands restrict populations of mosquitoes, but global climate change will likely result in more frequent El Niño events.[20] These bring more moisture to the islands and could cause more mosquitoes and mosquito-borne diseases to occur. It may be a matter of *when*, rather than *if*, we will see a repeat of the devastation that malaria caused to the birds of the Hawaiian Islands.

An example of an introduced parasite already having an effect in Galapagos is a botfly, *Philornis downsi*.[21] Botflies had not been recorded in Galapagos before the 1960s. Female flies lay eggs in nests. First-stage larvae feed on blood inside the nostrils of baby birds, and older larvae feed on blood by scratching wounds in the chicks' skin. Up to a hundred larvae can occur in one nest. Currently, bots occur in 97% of nests of the following species on Santa Cruz Island: yellow warblers (*Dendroica petechia*), mockingbirds (*Nesomimus parvulus*), dark-billed cuckoos (*Coccyzus melacoryphus*), smooth-billed anis (*Crotophaga ani*), seven species of Darwin's finches (species of *Geospiza* and *Camarhynchus*), and two species of flycatchers (*Pyrocephalus rubinus* and *Myriarchus magnirostris*).

Birds become infested regardless of the type and location of their nests. In an experiment in which nests of small ground finches (*Geospiza fuliginosa*) and medium ground finches (*G. fortis*) were treated with an insecticide (pyrethrin at a concentration of 1%), average botfly numbers were reduced from 21 per untreated nest to less than 1 per treated nest.[22] Chicks from treated nests gained more mass than parasitized birds and had higher hemoglobin levels. Additionally, fledging success improved by more than two-fold in treated nests.

Botflies are having a serious impact on Galapagos birds. The flies are widely distributed on all but the smallest, most arid islands (such as Genovesa and Daphne). Their drastic effect on fledging success will be amplified for birds that usually have small clutches, like tree finches (various species of *Camarhynchus*). Warbler finches (*Certhidea fusca*) from Floreana Island may already be extinct—perhaps because of *Philornis*. However, there may be some good news.

Researchers have found a parasitic wasp that attacks botflies. Careful trials will need to be done before conservationists would ever think of releasing these on Galapagos, but if they are host-specific enough, they may be employed as a biological control. However, one remedy is already being employed. Sarah Knutie and her colleagues have induced Darwin's finches to self-fumigate their nests with cotton fibers impregnated with permethrin insecticide.[23] Knutie put out 30 cotton dispensers containing cotton treated with permethrin on Santa Cruz Island, in addition to supplying cotton treated with water for purposes of comparison. Finches readily incorporated the cotton, at a rate of 85% among 26 active nests, resulting in 13 nests with permethrin cotton and 9 with untreated cotton. Nests with the insecticide had an average of 15 parasites, whereas the nests with untreated cotton had more than twice that number. In the 8 nests that had a gram or more of permethrin cotton, no parasites were found. The insecticide-containing nests resulted in increased fledging success. This research, in which birds treat themselves, shows that sometimes we can find clever ways of solving conservation problems.

To its credit, the government of Ecuador is taking measures to try to protect the wonderful and unique birds of Galapagos, along with its other wildlife. Visitor luggage and cargo from the mainland are inspected to try to prevent the accidental introduction of foreign organisms. There are also limits on the number of visitors allowed on the islands. Some introduced animals, like pigeons (*Columba livia*) and goats, are being eradicated from the islands, but it is economically, socially, and politically impossible to eliminate chickens.

Only young chicks that have not been vaccinated may be imported from aviculture companies on the mainland. Unfortunately, these facilities do not always follow strict biosecurity protocols. Continuous monitoring of domestic and wild animals by veterinary pathologists, along with surveys by biologists, must be used to detect any new disease problems—termed "pathogen pollution"—before these cause a mass die-off.

This work has already begun—more than 3,000 birds from 17 species, collected on 13 islands have been sampled to get baseline parasite data.[24] One interesting finding from this study is that, like their hosts, parasites on the Galapagos have diversified over evolutionary time. New species of lice and mites have been discovered, and very likely, more will be discovered as work continues. This parasite survey has also led to some basic knowledge about hosts. A feather louse (*Degeeriella regalis*), found only on Galapagos hawks (*Buteo galapagoensis*) on the islands, also infests these hawks' closest relatives on the mainland, Swainson's hawks (*B. swainsoni*). This louse is transmitted from parents to offspring. Lice mitochondrial DNA has shown that Galapagos hawks split from their common ancestor with Swainson's hawks less than 200,000 years ago, likely making Galapagos hawks the youngest endemic species of bird in the islands.

Conservation ecologists and Galapagos National Park administrators must develop models and contingency plans to efficiently deal with new disease outbreaks. Despite these efforts, it seems inevitable that the flying zoo in the Galapagos will drastically change due to human activity—habitat destruction, pollution, or accidental introduction of novel parasites will cause disruptions.

POLLUTION AND ITS EFFECTS

Globally, pollution also affects the flying zoo. Organochlorines (ocs) are persistent organic molecules that have been used for industrial and agricultural purposes—the infamous DDT is among these molecules. ocs accumulate in fatty tissues of animals, do not break down, and become concentrated in predators that consume prey containing them, a phenomenon called biomagnification.

Organochlorines have made their way by wind and water currents into regions like the Arctic, which are very remote from the places where these chemicals are made or used. Thanks to Rachel Carson and her book *Silent Spring*,[25] many ocs have been banned in large parts of the world due to their serious negative effects on bird reproduction, bird populations, and their potential to contaminate humans. Besides affecting reproduction, ocs can cause neurological problems and probably also suppress immune functions.

In 2000, Kjetil Sagerup and his colleagues collected forty adult glaucous gulls (*Larus hyperboreus*) from Bear Island in the Barents Sea north of Norway.[26] Glaucous gulls are large, generalist scavenger-predators. They have a circumpolar distribution and nest in Greenland, Iceland, Russia, and the Canadian Arctic. The birds were examined for gastrointestinal helminth parasites (nematodes, flukes, tapeworms, and spiny-headed worms), and samples of each bird's liver were taken to determine levels of ocs. The researchers found a positive correlation between the numbers of nematode parasites present in the gulls and the levels of ocs in their livers.

The life cycles of the nematodes were not consistent with the idea that high oc levels were due to gulls eating a particular type of intermediate host—that is, the relationship was not due to diet. The most likely explanation for the correlation was that ocs were compromising the birds' immune systems, leading to greater numbers of worms. Unfortunately, the researchers did not attempt to measure immunocompetence, so this relationship is still speculative. And an interesting twist on this problem was revealed in an experiment involving glaucous gulls by Jan Bustnes and colleagues.[27] After treating birds with anti-parasite medications, they found that gulls were better able to deal with pollutants. Birds that deal with natural stressors such as parasites seem to have trouble dealing with unnatural, human-induced stress such as pollutants.

Regardless of the exact nature of the relationship between parasites and pollution, this relationship is becoming an increasingly important factor to consider for bird conservation, because parasites can reduce bird populations—and the effects of pollution may be far-reaching, as suggested by the case of glaucous gulls in the remote Arctic.

The production and use of persistent pesticides has been reduced since the 1960s. However, these chemicals have been replaced by ones that break down quickly but are much more toxic. Inappropriate use of these agents directly and indiscriminately kill many birds that are attracted to dead and dying insects. Farmers—who in less-developed countries are often illiterate—are also in great danger when they use these pesticides.[28]

HABITAT PRESSURES

In addition to the problems of chemical and pathogen pollution, our expanding population demands more and more land. Natural wild areas (including some of the most remote places on Earth, like the Amazon Basin) are being deforested, drained, flooded, and carved up for agriculture, mining, or suburban construction. Even activities that by themselves would not seem to involve much habitat destruction can result in major environmental impacts, as was the case in Yasuni National Park in eastern Ecuador, where oil had been discovered.

When forests are exploited in the most environmentally sensitive way possible, oil wells and pump jacks use very small parcels of natural tropical rainforest; however, exploring, building, and servicing these sites demand that roads, bridges, and cut lines be built. Later, pipelines are constructed to carry oil to refineries, usually far from extraction sites. With roads come hunters, poachers, settlers, small farms, and later towns, with ever more development—all with the encouragement of governments in the name of economic growth. This onslaught continues despite existing laws designed to protect natural areas or the land rights of Indigenous peoples. In Ecuador, land was divided and parceled out as oil leases before the creation of Yasuni Park, so resource companies—which are mostly foreign-owned—can legally explore and destroy the national park.

This pattern is not unique to Ecuador. It has happened to rainforests and boreal forests all over the world. Huge continuous tracts of mostly pristine, old-growth forests get dissected into smaller and smaller parcels for immediate and short-term economic benefit—an ecological trend called habitat fragmentation. With short-term gains, however, often come long-term pains. The flying zoo is proving this now.

Songbirds live in two worlds. From April to September they feed, court, reproduce, and care for young in coniferous and deciduous forests, fields, and backyards in North America. From October to March, they defend territories, live, and feed in the tropical forests of the Caribbean Islands and Central and South America. Habitat fragmentation is causing alarming declines in the populations of many types of songbirds.[29]

Many factors are contributing to the reduced number of songbirds, including pesticide pollution and drastic fluctuations in climate patterns with frequent severe storms. But habitat loss due to fragmentation, both in northern and tropical forests, is the main problem.

Exactly how habitat fragmentation affects the flying zoo is not known. In some cases, reduced host populations can result in a decline in parasites because the chain of transmission is weakened. But sometimes, habitat fragmentation forces birds into substandard territories, causing poorer nutrition, poorer immune function, and greater transmission of parasites. For instance, Kirtland's warblers (*Setophaga kirtlandii*) have been driven to the brink of extinction, and one proximate reason for their decline is parasitism by brown-headed cowbirds, which has been facilitated by forest fragmentation.[30] Conservationists now monitor warbler nests and remove cowbird eggs to save the warblers. The tangle of threads that intertwine in the relationship between birds and their parasites limits our ability to accurately predict the outcomes of human-caused environmental change.

This complexity is highlighted by the story of Lyme disease, which involves interactions between habitat fragmentation, deer, and the now-extinct passenger pigeon. Lyme disease is caused by the bacterium *Borrelia burgdorferi*. It is transmitted by ticks (*Ixodes scapularis* and *I. pacificus*) that use small mammals or ground-dwelling birds as first hosts and larger mammals, typically white-tailed deer (*Odocoileus virginianus*), as second hosts. When a person is bitten by an infected tick, Lyme disease can result. The bacteria spread from bite sites and cause rashes, arthritis, and eventually, if untreated, neurological problems. Deforestation and habitat fragmentation can amplify the number and diversity of ground-dwelling birds and rodents. This, in turn, likely facilitates increased cases of Lyme disease.[31] However, the full story is likely to be more complicated than this.

An intriguing possibility is that the extinction of passenger pigeons (*Ectopistes migratorius*), the last of which died on September 1, 1914, may have promoted the spread of Lyme disease in North America. Before European settlement in North America, passenger pigeons were probably not as numerous as historical accounts report—there is little evidence of feathers or bones of passenger pigeons at Indigenous settlements, even though the remains of other small birds and mammals have been found. As settlers later discovered, the birds were tasty and easy to kill, so it does not make sense that Indigenous peoples would have not eaten them.

Europeans probably reduced deer numbers when they cleared old-growth forests, converted land to agriculture, and hunted. As a result, the population of passenger pigeons exploded, because passenger pigeons probably ate seeds (including acorns) that also sustained mice and deer. Less deer meant more food for the passenger pigeons. Most of us have heard the famous story of John James Audubon (1785–1851) who reported a flock of passenger pigeons that flew continuously overhead for three days! However, settlers soon exterminated the easily killed birds.

When passenger pigeons were exterminated, this may have removed a major competitor of deer. Now it was the tick-carrying deer population that exploded. As most of us who live in cities know, the population of white-tailed deer has grown in recent years; they are now very common urban wildlife. This situation sets the stage for a large-scale invasion of Lyme disease, which was probably brought to North America by European settlers. In addition, climate change seems to be associated with an increase in tick populations and an expansion of their northern range.[32] The link between passenger pigeons and Lyme disease is based on a lot of speculation and we don't yet know how accurate it is. But it does make us consider how the exploitation and manipulation of nature might affect parasites and, ultimately, human well-being.

How might parasites fare with coming environmental changes? Although some pathogens, like malaria, may expand their populations, Carrie Cizauskas and colleagues propose that most parasites, and particularly helminths and arthropods, will become more vulnerable to extinction.[33] They argue that many will go extinct as their hosts go extinct, through the process of co-extinction. Some parasites may even go extinct before their host is extinct due to a decline of their host population. For example, *Colpocephalum californici* feather lice are now extinct, but a remnant population of their hosts, California condors (*Gymnogyps californianus*), still exists.[34]

Several factors likely affect parasite survival and, in many cases, make parasites even more vulnerable than free-living species. For example, parasites that rely on "cold-blooded" species such as reptiles may be at especially high risk; because these hosts cannot control their body temperatures, climate changes will likely drastically affect their population sizes and ranges. Other factors that may endanger parasites include host specificity (more specific parasites have smaller populations and less genetic diversity), life cycles (stages outside a bird face volatile environments and intermediate hosts may suffer extinction), and host body size (parasites of larger hosts have more internal niches). Thus, theoretically, bird parasites will be highly vulnerable to extinction.

Perhaps many of you are thinking that this can only be good news for birds, but one forceful lesson that arises from our understanding of the flying zoo is that changes to biodiversity often have unexpected consequences. The parasite–bird relationship is far from simple. Certain parasitic interactions can become mutualistic. Some normal parasites in hosts can reduce opportunities for more pathogenic ones. Moreover, parasitized intermediate hosts may be important in feeding ecology. Thus, parasites can affect the dynamics and stability of entire communities.

Author and inventor Buckminster Fuller coined the term *synergy* for systems where interactions between well-known and understood parts (like the springs and gears of a clock) cannot be used to predict the consequences of the whole system (telling time). Certainly, the interactions between humans, the

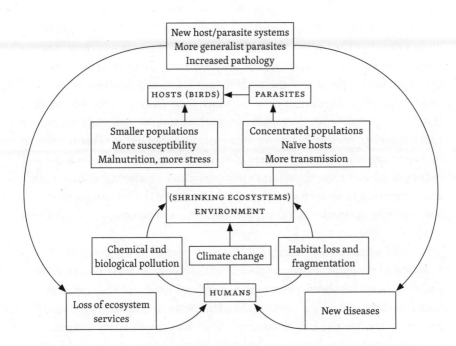

FIGURE 10.1 A model of the relationships between humans, the environment, and the flying zoo.

environment, and the flying zoo are synergistic. A general model was devised by John Holmes,[35] and I have adapted it here as a conclusion in Figure 10.1.

We are changing the natural environment by destroying and fragmenting habitat, by altering climate, and by polluting the Earth with chemicals and with introduced organisms. We know that this is shrinking and destabilizing ecosystems, particularly in tropical rainforests, the Arctic, coastal estuaries, wetlands, temperate old-growth forests and boreal forests. This results in greater concentrations of birds and displacement into marginal habitats. Increased stress and close contact between birds of the same and different species foster the spread of diseases among birds. Contact with immunologically naïve birds is especially devastating.

Parasites, which in their normal host have co-evolved to be mostly benign, may become dangerous pathogens because of increased transmission. Generalist parasites—those that can live at least temporarily in a variety of hosts, but tend to be concentrated in one—become more common in fragmented ecosystems. Unfortunately, these parasites have not co-evolved with most of the birds they contact, so there has not been enough selection pressure or time to reduce their pathogenicity. In concentrated host populations, microparasites such as viruses may undergo genetic recombination, resulting in new pathogens. This scenario is of great concern in Asia today, where large numbers of birds are kept in dense populations. If influenza viruses recombine with virus particles in mammals such as pigs, then migrating birds may spread the new virus into the human population. The resulting epizootic—a human epidemic due to animals—could kill thousands of people.

Sadly, negative anthropogenic effects on wild birds affects us all. Ecosystem loss equals more and new diseases like West Nile and Lyme. In addition, as we lose birds, we lose the ecosystem functions they perform for us, such as controlling insect and rodent pests, and pollinating plants. And of course, we also lose in other ways: we and our children will miss the chance to see and hear these beautiful jewels of the natural world. In a decade, I hope that geese will still undergo their annual migration and still tell me that spring is here.

Other than ecologists and climatologists, few of us have paid attention to the unnatural changes that happened in the past and that are happening today. But there is reason to hope. There are simple things all of us can do to stop and even reverse the dreadful trends that we are now seeing. Indeed, the times they are a-changin'—but if we take responsibility as caretakers of our planet, we can still save the flying zoo, the Earth, and ourselves.

TABLE 10.1.   Ways we can help save the Earth and preserve biodiversity

THINGS TO DO	WAYS TO DO IT	BENEFITS
Reduce use of fossil fuels.	Buy smaller, more fuel-efficient vehicles. Use mass transit or bicycles or walk whenever possible. Use energy-efficient appliances.	Reduces pollution of air and water. Helps to reverse climate change.
Reduce use of water.	Use low-flow toilets and "waterless" water heaters. Plant native, drought-resistant plants. Drink tap water instead of bottled water.	Preserves wetlands. Provides food for native wildlife species. Saves energy.
Use local products.	Buy produce from farmers' markets.	Reduces energy. Stimulates local economy. Reduces habitat fragmentation and water use in other ecosystems.
Buy certified organic produce and shade-grown coffee.	Read product labels.	Reduces the use of pesticides. Reduces the use of fertilizers made from fossil fuels. Stimulates the local economy. Saves rainforest and cloud forest ecosystems.
Landscape with native plants, especially grasses, trees, and shrubs.	Buy seeds or plants from local producers.	Provides food and habitat for native wildlife. Reduces introduction of invasive species.
Get in touch with and introduce children to nature.	Observe and identify plants, fungi, and animals in your own backyard. Visit local natural areas. Join nature clubs. Buy field guides for local wildlife.	Provides a sense of well-being and relaxation. Stimulates our curiosity. Helps us to feel connected with other living things.
Keep aware and don't despair.	Read critically. Hold politicians accountable. Vote.	You can make a difference!

NOTES

1   A WORLD ON A BIRD

1. Clayton, D.H., and Walther, B.A. (2001). Influence of host ecology and morphology on diversity of Neotropical bird lice. *Oikos* 94, 455–467.
2. Moller, A.P. (1991). Parasites, sexual ornaments, and mate choice in the barn swallow. In *Bird-parasite interactions: Ecology, evolution, and behaviour.* Loye, J.E., and Zuk, M. (Eds.), 328–343. Oxford: Oxford University Press.
3. Kose, M., Mandi, R., and Moller, A.P. (1999). Sexual selection for white tail spots in the barn swallow in relation to habitat choice by feather lice. *Animal Behaviour* 58, 1201–1205.
4. Walther, B.A., and Clayton, D.H. (2004). Elaborate ornaments are costly to maintain: Evidence for high maintenance handicaps. *Behavioral Ecology* 16, 89–95.
5. Bush, S.E., Kim, D., Reed, M., and Clayton, D.H. (2010). Evolution of cryptic coloration in ectoparasites. *American Naturalist* 176, 529–535.
6. Clay, T. (1949). Some problems in the evolution of a group of ectoparasites. *Evolution* 3, 279–299.
7. Kozlovsky, D.Y., Brown, S.L., Branch, C.L., Roth II, T.C., and Pravosudov, V.V. (2014). Chickadees with bigger brains have smaller digestive tracts: A multipopulation comparison. *Brain, Behavior, and Evolution* 84, 172–180.
8. Cuthill, I.C., Allen, W.L., Arbuckle, K., Caspers, B., Chaplin, G., Hauber, M.E., Hill, G.E., Jablonski, N.G., Jiggins, C.D., Kelber, A., Mappes, J., Marshall, J., Merrill, R., Osorio, D., Prum, R.O., Roberts, N.W., Roulin, A., Rowland, H.M., Sherratt, T.N., Skelhorn, J., Speed, M.P., Stevens, M., Stoddard, M.C., Stuart-Fox, D., Talas, L., Tibbetts, E., and Caro, T. (2017). The biology of color. *Science* 357, 1–29. *Also:* Prum, R.O., Torres, R.H., Williamson S., and Dyck, J. (1998). Coherent light scattering by blue feather barbs. *Nature* 396, 28–29.
9. Doucet, S.M., and Montgomerie, R. (2003). Multiple sexual ornaments in satin bowerbirds: Ultraviolet plumage and bowers signal different aspects of male quality. *Behavioral Ecology* 14, 503–509.
10. Badas, E.P., Martinez, J., Rivero-de Aguilar, J., Ponce, C., Stevens, M., and Merino, S. (2018). Colour change in a structural ornament is related to individual quality, parasites, and mating patterns in the blue tit. *Science of Nature* 105, 1–17.
11. Boyd, E.M., Diminno, R.L., and Nesslinger, C. (1956). Metazoan parasites of the blue jay, *Cyanocitta cristata* L. *Journal of Parasitology* 42, 332–346. *Also:* Cooper, C.L., and Crites, J.L. (1974). The helminth parasites of the blue jay, *Cyanocitta cristata bromia,* from South Bass Island, Ohio. *Canadian Journal of Zoology* 52, 1421–1423.
12. Zirpoli, J.A., Black, J.M., and Gabriel, P.O. (2013). Parasites and plumage in Steller's jays: An experimental field test of the parasite-mediated handicap hypothesis. *Ethology, Ecology, and Evolution* 25, 103–116.

13. Mennill, D.J., Doucet, S., Montgomerie, M.R., and Ratcliffe, L.M. (2003). Achromatic color variation in black-capped chickadees, *Poecile atricapillar*: Black and white signals of sex and rank. *Behavioral Ecology and Sociobiology* 53, 350-357.

14. D'Alba, L., Van Hemert, C., Handel, C.M., and Shawkey, M.D. (2011). A natural experiment on the condition-dependence of achromatic plumage reflectance in black-capped chickadees. *PLoS One* 6, 1-8.

15. Wilkinson, L.C., Handel, C.M., Van Hemert, C., Loiseau, C., and Sehgal, R.N.M. (2016). Avian malaria in a boreal resident species: Long-term temporal variability and increased prevalence in birds with avian keratin disorder. *International Journal of Parasitology* 46, 281-290.

16. Handel, C.M., Pajot, L.M., Matsuoka, S.M., Van Hemert, C., Tterenzi, J., Talbot, S.L., Mulcahy, D.M., Meteyer, C.U., and Trust, K.A. (2010). Epizootic of beak deformities among wild birds in Alaska: An emerging disease in North America. *The Auk* 127, 882-898.

17. D'Alba, L., Van Hemert, C., Handel, C.M., and Shawkey, M.D. (2011). A natural experiment on the condition-dependence of achromatic plumage reflectance in black-capped chickadees. *PLoS One* 6, 1-8.

2   LICE
It's a Beautiful Life

1. Barker, S.C. (1994). Phylogeny and classification, origins, and evolution of host associations of lice. *International Journal of Parasitology* 24, 1285-1201.

2. Wappler, T., Smith, V.S., and Dalgleish, R.C. (2004). Scratching an ancient itch: An Eocene bird louse fossil. *Proceedings of the Royal Society of London. Series B: Biological Sciences* 271, 255-258.

3. Grimaldi, D., and Engel, M.S. (2006). Fossil Liposcelididae and the lice ages (Insecta: Psocodea). *Proceedings of the Royal Society of London. Series B: Biological Sciences* 273, 625-633.

4. Rasnitsyn, A.P., and Zherikhin V.V. (1999). First fossil chewing louse from the lower Cretaceous of Baissa, Transbaikalia (Insecta, Pediculida = Phthiriaptera, Saurodectidae fam. N.). *Russian Entomological Journal* 8, 253-255.

5. Dalgleish, R.C., Palma, R.L., Price, R.D., and Smith, V.S. (2006). Fossil lice (Insecta: Phthiraptera) reconsidered. *Systematic Entomology* 31, 648-651.

6. Marshall, A.G. (1981). *The ecology of ectoparasitic insects.* New York: Academic Press.

7. Futuyma, D.J., and Slatkin, M. (Eds). (1983). *Coevolution.* Sunderland, MA: Sinauer.

8. Noble, E.R., and Noble, G.A. (1973). *Parasitology: The biology of animal parasites (3rd ed.).* Philadelphia: Lea and Febiger.

9. Clayton, D.H., and Price, R.D. (1999). Taxonomy of New World *Columbicola* (Phthiraptera: Philopteridae) from the Columbiformes (Aves), with descriptions of five new species. *Annals of the Entomological Society of America* 92, 675-685.

10. Clayton, D.H., and Johnson, K.P. (2003). Linking coevolutionary history to ecological process: Doves and lice. *Evolution* 57, 2335-2341.

11. Bartlow, A.W., Villa, S.M., Thompson, M.W., and Bush, S.E. (2016). Walk or ride? Phoretic behavior of amblyceran and ischnoceran lice. *International Journal of Parasitology* 46, 221-227.

12. Paterson, A.M., and Gray, R.D. (1997). Host-parasite co-speciation, host switching, and missing the boat. In *Host-parasite evolution: General principles and avian models*. Clayton, D.H., and Moore, J. (Eds), 236-250. Oxford: Oxford University Press.

13. Clayton, D.H. (1990). Host specificity of *Strigophilus* owl lice (Ischnocera: Philopteridae), with the description of new species and host associations. *Journal of Medical Entomology* 27, 257-265.

14. Hahn, D.C., Price, R.D., and Osenton, P.C. (2000). Use of lice to identify cowbird hosts. *The Auk* 117, 943-951.

15. Rothschild, M., and Clay, T. (1952). *Fleas, flukes and cuckoos: A study of bird parasites*. London, UK: Philosophical Library.

16. Feduccia, A. (1999). *Origin and evolution of birds (2nd ed.)*. New Haven, CT: Yale University Press.

17. Peters, D.A. (1987). Juncitarsus merkeli n. sp. stützt die Ableitung der Flamingos von Regenpfeifervögeln (Aves: Charadriiformes: Phoenicopteridae). *Courier Forschungsinstitut Senckenberg* 9, 141-155.

18. Mayr, G. (2004). Morphological evidence for sister group relationship between flamingos (Aves: Phoenicopteridae) and grebes (Podicipedidae). *Zoological Journal of the Linnean Society* 140, 157-169.

19. Johnson, K.P., Kennedy, M., and McCracken, K.G. (2006). Reinterpreting the origins of flamingo lice: Cospeciation or host-switching? *Biological Letters* 2, 275-278.

20. Booth, D.T., Clayton, D.H., and Block, B.A. (1993). Experimental demonstration of the energetic costs of parasitism in free-ranging hosts. *Proceedings of the Royal Society of London. Series B: Biological Sciences* 253, 125-129.

21. Cotgreave, P., and Clayton, D.H. (1994). Comparative analysis of time spent grooming by birds in relation to parasite load. *Behaviour* 131, 171-187.

22. Clayton, D.H., and Cotgreave, P. (1994). Relationship of bill morphology to grooming behaviour in birds. *Animal Behaviour* 47, 195-201.

23. Clayton, D.H., and Walther, B.A. (2001). Influence of host ecology and morphology on the diversity of Neotropical bird lice. *Oikos* 94, 455-467.

24. Rozsa, L. (1993). An experimental test of the site specificity of preening to control lice in feral pigeons. *Journal of Parasitology* 79, 968-970.

25. Samuel, W.M., Williams, E.S., and Rippin, A.B. (1982). Infestations of *Piagetiella peralis* (Mallophaga: Menoponidae) on juvenile white pelicans. *Canadian Journal of Zoology* 60, 951-953.

26. Rothschild, M., and Clay, T. (1952). *Fleas, flukes and cuckoos: A study of bird parasites*. London, UK: Philosophical Library.

27. Bartlett, C.M., and Anderson, R.C. (1989). Mallophagan vectors and the avian filaroids: New subspecies of *Pelecitus fulicaeatrae* (Nematoda: Filaroidea) in sympatric North American hosts, with development, epizootiology, and pathogenesis of the parasite in *Fulica americana* (Aves). *Canadian Journal of Zoology* 67, 2821-2833.

28. Bush, S.E., Kim, D., Reed, M., and Clayton, D.H. (2010). Evolution of cryptic coloration in ectoparasites. *The American Naturalist* 176, 529-535.

29. Johnson, K.P., Bush, S.E., and Clayton, D.H. (2005). Correlated evolution of host and parasite body size: tests of Harrison's rule using birds and lice. *Evolution* 59, 1744-1753.

30. Clayton, D.H., Bush, S.E., and Goates, B.M. (2003). Host defense reinforces host-parasite cospeciation. *Proceedings of the National Academy of Sciences* 100, 15694-15699.

31. Nelson, B.C., and Murray, M.D. (1971). The distribution of mallophaga on the domestic pigeon *Columba livia*. *International Journal of Parasitology* 1, 21-29.

32. Clayton, D.H., Koop, J.A.H., Harbison, C.W., Moyer, B.R., and Bush, S.E. (2010). How birds combat ectoparasites. *The Open Ornithology Journal* 3, 41-71.

33. Suarez-Rodriguez, M., and Garcia, C.M. (2017). An experimental demonstration that house finches add cigarette butts in response to ectoparasites. *Journal of Avian Biology* 48, 1316-1321.

34. Dumbacher, J.P. (1999). Evolution and toxicity in pitohuis: I. Effects of homobatrachotoxin on chewing lice (order Phthiraptera). *The Auk* 116, 957-963.

35. Lovette, I.J., and Fitzpatrick, J.W. (2016). *The Cornell Lab of Ornithology handbook of bird biology, 3rd ed.* Chichester, UK: Wiley.

36. Clayton, D.H., Gregory, R.D., and Price, R.D. (1992). Comparative ecology of Neotropical bird lice (Insecta: Phthiraptera). *Journal of Animal Ecology* 61, 781-795.

37. Moller, A.P., and Rozsa, L. (2005). Parasite biodiversity and host defenses: Chewing lice and immune response of their avian hosts. *Oecologia* 142, 169-176.

38. Whiteman, N.K., and Parker, P.G. (2004). Body condition and parasite load predict territory ownership in the Galapagos hawk. *Condor* 106, 915-921.

39. Rozsa, L., Rekasi, J., and Reiczigel, J. (1996). Relationship of host coloniality to the population ecology of avian lice (Insecta: Phthiraptera). *Journal of Animal Ecology* 65, 242-248.

40. Hillgarth, N. (1996). Ectoparasite transfer during mating in ring-necked pheasants *Phasianus colchicus*. *Journal of Avian Biology* 27, 260-262.

41. Whiteman, N.K., and Parker, P.G. (2004). Body condition and parasite load predict territory ownership in the Galapagos hawk. *Condor* 106, 915-921.

42. As reported in: Noble, E.R., and Noble, G.A. (1973). *Parasitology: The biology of animal parasites (3rd ed)*. Philadelphia: Lea and Febiger.

3   FLEAS
The Circus in the Zoo

1. Evans, I.H. (Ed) (1981). *Brewer's dictionary of phrase and fable*. New York: Harper and Row.

2. Whiting, M.F. (2002). Mecoptera is paraphyletic: Multiple genes and phylogeny of Mecoptera and Siphonaptera. *Zoologica scripta* 31, 93-104.

3. Traub, R., and Rothschild, M. (1983). Evolution of the Ceratophyllidae. In *The Rothschild collection of fleas. The Ceratophyllidae: Key to the genera and host relationships.* Traub, R., Rothschild, M., and Haddow, J.F. (Eds), 188-201. London, UK: Academic Press.

4. Traub, R., and Rothschild, M. (1983). Evolution of the Ceratophyllidae. In *The Rothschild collection of fleas. The Ceratophyllidae: Key to the genera and host relationships.* Traub, R., Rothschild, M., and Haddow, J.F. (Eds), 188-201. London, UK: Academic Press.

5. Amrine, J.W., and Jerabek, M.A. (1983). Possible ultrasonic receptors on fleas. *Annals of the Entomological Society of America* 76, 395-399.

6. Lehane, M.J. (1991). *Biology of blood-sucking insects*. London, UK: HarperCollins.

7. Traub, R., and Rothschild, M. (1983). Evolution of the Ceratophyllidae. In *The Rothschild collection of fleas. The Ceratophyllidae: Key to the genera and host relationships.* Traub, R., Rothschild, M., and Haddow, J.F. (Eds), 188-201. London, UK: Academic Press.
8. Marshall, A.G. (1981). *The ecology of ectoparasitic insects.* New York: Academic Press.
9. Tripet, F., Christe, P., and Moller, A.P. (2002). The importance of host spatial distribution for parasite specialization and speciation: A comparative study of bird fleas (Siphonaptera: Ceratophyllidae). *Journal of Animal Ecology* 71, 735-748.
10. Richner, H., Oppliger, A., and Christe, P. (1993). Effect of an ectoparasite on reproduction in great tits. *Journal of Animal Ecology* 62, 703-710.
11. Christe, P., Richner, H., and Oppliger, A. (1996). Begging, food provisioning, and nestling competition in great tit broods infested with ectoparasites. *Behavioral Ecology* 7, 127-131.
12. Christe, P., Richner, H., and Opplinger, A. (1996). Of great tits and fleas: Sleep baby sleep... *Animal Behavior* 52, 1087-1092.
13. Fitze, P.S., Tschirren, B., and Richner, H. (2004). Life history and fitness consequences of ectoparasites. *Journal of Animal Ecology* 73, 216-226.
14. Heeb, P., Werner, I., Kolliker, M., and Richner, H. (1998). Benefits of induced host responses against an ectoparasite. *Proceedings of the Royal Society of London. Series B: Biological Sciences* 265, 51-56.
15. Hurtrez-Bousses, S., Perret, P., Renaud, F., and Blondel, J. (1997). High blowfly parasitic loads affect breeding success on a Mediterranean population of blue tits. *Oecologia* 112, 514-517.
16. O'Brien, E.L., and Dawson, R.D. (2005). Perceived risk of ectoparasitism reduces primary reproductive investment in tree swallows *Tachycineta bicolor. Journal of Avian Biology* 36, 269-275.
17. Du Feu, C.R. (2012). To clean, or not to clean? That is the question. *Avian Biology Research* 5, 252-254.
18. Du Feu, C.R. (2012). To clean, or not to clean? That is the question. *Avian Biology Research* 5, 252-254.
19. Hart, B.L. (1997). Behavioural defense. In *Host-parasite evolution: General principles and avian models.* Clayton, D.H., and Moore, J. (Eds), 59-97. Oxford: Oxford University Press.
20. Buechler, K., Fitze, P.S., Gottstein, B., Jacot, A., and Richner, H. (2002). Parasite-induced maternal response in a natural bird population. *Journal of Animal Ecology* 71, 247-252.
21. Postma, E., Siitari, H., Schwabl, H., Richner, H., and Tschirren, B. (2014). The multivariate egg: Quantifying within—and among—clutch correlations between maternally derived yolk immunoglobulins and yolk androgens using multivariate mixed models. *Oecologia* 174, 631-638.
22. Moller, A.P., Christe, P., and Garamszegi, Z. (2004). Coevolutionary arms races: Increased host defense promotes specialization by avian fleas. *Journal of Evolutionary Biology* 18, 46-59.
23. Tripet, F., and Richner, H. (1997). The coevolutionary potential of a "generalist" parasite, the hen flea *Ceratophyllus gallinae. Parasitology* 115, 419-427.
24. Rothschild, M., and Clay, T. (1952). *Fleas, flukes, and cuckoos.* London, UK: Collins.
25. Jorink, E. (2010). *Reading the book of nature in the Dutch Golden Age, 1575-1715.* Leiden: Koninklujke Brill.

1. Sonenshine, D.E., and Roe, R.M. (Eds). (2014). *Biology of ticks, 2nd ed., vols. I and II.* Oxford: Oxford University Press.

2. Ribeiro, J., Alcaron-Chaidez, F., Francischetti, I.M.B., Mans, B., Mather, T.N., Valenzuela, J.G., and Wikel, S.K. (2006). An annotated catalog of salivary gland transcripts from *Ixodes scapularis* ticks. *Insect Biochemistry and Molecular Biology* 36, 111–129.

3. Kaufman, W.R. (1989). Tick-host interaction: A synthesis of current concepts. *Parasitology Today* 5, 47–56.

4. Frenot, Y., de Oliveira, E., Gauthier-Clerc, M., Deunff, J., Bellido, A., and Vernon, P. (2001). Life cycle of the tick *Ixodes uriae* in penguin colonies: Relationships with host breeding activity. *International Journal of Parasitology* 31, 1040–1047.

5. Davis, L.S., and Renner, M. (2003). *Penguins.* New Haven, CT: Yale University Press. *Also:* Williams, T.D. (1995). *The penguins.* Oxford: Oxford University Press.

6. Frenot, Y., de Oliveira, E., Gauthier-Clerc, M., Deunff, J., Bellido, A., and Vernon, P. (2001). Life cycle of the tick *Ixodes uriae* in penguin colonies: Relationships with host breeding activity. *International Journal of Parasitology* 31, 1040–1047.

7. McCoy, K.D., and Tirard, C. (2002). Reproductive strategies of the seabird tick *Ixodes uriae* (Acari: Ixodidae). *Journal of Parasitology* 88, 813–816.

8. Brooke, M. de L. (1985). The effect of allopreeening on tick burdens of molting Eudyptid penguins. *The Auk* 102, 891–892.

9. Murray, M.D., and Vestjens, W.J.M. (1967). Studies on the ectoparasites of seals and penguins III. The distribution of the tick *Ixodes uriae* White and the flea *Parapsyllus magellanicus heardi* de Meillon on Macquarie Island. *Australian Journal of Zoology* 15, 715–725.

10. Frenot, Y., de Oliveira, E., Gauthier-Clerc, M., Deunff, J., Bellido, A., and Vernon, P. (2001). Life cycle of the tick *Ixodes uriae* in penguin colonies: Relationships with host breeding activity. *International Journal of Parasitology* 31, 1040–1047.

11. Klompen, H., and Grimaldi, D. (2001). First Mesozoic record of a parasitiform mite: A larval argasid tick in Cretaceous amber (Acari: Ixodida: Argasidae). *Annals of the Entomological Society of America* 94, 10–15.

12. Wikel, S.K. (2014). Tick-host interactions. In *Biology of ticks, 2nd ed., vol. II.* Sonenshine, D.E., and Roe, R.M. (Eds), 88–128. Oxford: Oxford University Press.

13. Gauthier-Clerc, M., Jaulhac, B., Frenot, Y., Bachelard, C., Monteil, H., Le Maho, Y., and Handrich, Y. (1999). Prevalence of *Borrelia burgdorferi* (the Lyme disease agent) antibodies in King penguin *Aptenodytes patagonicus* in Crozet Archipelago. *Polar Biology* 22, 141–143.

14. Mangin, S., Gauthier-Clerc, M., Frenot, Y., Gendner, J.-P., and Le Maho, Y. (2003). Ticks *Ixodes uriae* and the breeding performance of a colonial seabird, king penguin *Aptenodytes patagonicus. Journal of Avian Biology* 34, 30–34.

15. Wikel, S.K. (2014). Tick-host interactions. In *Biology of ticks, 2nd ed., vol. II.* Sonenshine, D.E., and Roe, R.M. (Eds), 88–128. Oxford: Oxford University Press.

16. Gray, J.S., Estrada-Pena, A., and Vial, L. (2014). Ecology of nidicolous ticks. In *Biology of ticks, 2nd ed., vol. II.* Sonenshine, D.E., and Roe, R.M. (Eds), 3–38. Oxford: Oxford University Press.

17. Khalil, G.M. (1976). The subgenus *Persicargas* (Ixodoidea: Argasidae: *Argas*), 26. *Argas* (*P.*) *arboreus*: Effect of photoperiod on diapause induction and termination. *Experimental Parasitology* 40, 232–237.

18. Sonenshine, D.E. (1993). *Biology of ticks, vol. II*. Oxford: Oxford University Press.

19. Duffy, D.C., and Daturi, A. (1987). Diel rhythms of tick parasitism on incubating African penguins. *Medical and Veterinary Entomology* 1, 103–106.

20. Yunker, C. (1975). Tick-borne viruses associated with seabirds in North America and related islands. *Medical Biology* 53, 302–311.

21. Duffy, D.C. (1983). The ecology of tick parasitism on densely nesting Peruvian seabirds. *Ecology* 64, 110–119.

22. Duffy, D.C. (1980). *Comparative reproductive behavior and population regulation of seabirds of the Peruvian coastal current*. PHD dissertation, Princeton University.

5   MITES
Little Things Mean a Lot

1. Walter, D.E., and Proctor, H.C. (2013). *Mites: Ecology, evolution and behaviour. Life at a microscale, 2nd ed.* New York: Springer.

2. Blanco, G., Tella, J.L., and Potti, J. (2001). Feather mites on birds: Costs of parasitism or conditional outcomes? *Journal of Avian Biology* 32, 271–274.

3. Colwell, R.K. (1995). Effects of nectar consumption by the hummingbird flower mite *Proctolaelaps kirmsei* on nectar availability in *Hamelia patens*. *Biotropica* 27, 206–217.

4. Dabert, J., and Miranov, S.V. (1999). Origin and evolution of feather mites (Astigmata). *Experimental and Applied Acarology* 23, 437–454.

5. Burtt, E.H. Jr., Chow, W., and Babbitt, G.A. (1991). Occurrence and demography of mites of tree swallow, house wren, and eastern bluebird nests. In *Bird-parasite interactions: Ecology, evolution and behaviour*. Loye, J.E., and Zuk, M. (Eds), 104–122. Oxford: Oxford University Press.

6. Clark, L. (1991). The nest protection hypothesis: The adaptive use of plant secondary compounds by European starlings. In *Bird-parasite interactions: Ecology, evolution and behaviour*. Loye, J.E., and Zuk, M. (Eds), 205–221. Oxford: Oxford University Press.

7. Mullens, B.A., Hinkle, N.C., and Szijj, C.E. (2000). Monitoring northern fowl mites (Acari: Macronyssidae) in caged laying hens: Feasibility of an egg-based sampling system. *Journal of Economic Entomology* 93, 1045–1054.

8. Owen, J.B., and Mullens, B.A. (2004). Influence of heat and vibration on the movement of the northern fowl mite (Acari: Macronyssidae). *Journal of Medical Entomology* 41, 865–872.

9. Valera, F., Casa-Criville, A., and Hoi, H. (2003). Interspecific parasite exchange in a mixed colony of birds. *Journal of Parasitology* 89, 245–250.

10. Latta, S.C. (2003). Effects of scaly-leg mite infestations on body condition and site fidelity of migratory warblers in the Dominican Republic. *The Auk* 120, 730–743.

11. Benckman, C.W., Colquitt, J.S., Gould, W.R., Fetz, T., Keenan, P.C., and Santisteban, L. (2005). Can selection by an ectoparasite drive a population of red crossbills from its adaptive peak? *Evolution* 59, 2025–2032.

12. Pence, D.B. (1979). Congruent inter-relationships of the rhinonyssinae (Dermanyssidae) with their avian hosts. *Recent Advances in Acarology* 2, 371–377.

13. Fain, A. (1969). Adaptation to parasitism in mites. *Acarologia* 11, 429–449.

14. Bell, P.J. (1996). The life history and transmission biology of *Sternostoma tracheacolum* Lawrence (Acari: Rhinonyssidae) associated with the Gouldian finch *Erythrura gouldiae*. *Experimental and Applied Acarology* 20, 323–324.

15. Hendrix, C.M., Kwapien, R.P., and Porch, J.R. (1987). Visceral and subcutaneous acariasis caused by hypopi of *Hypodectes propus bulbuci* in the cattle egret. *Journal of Wildlife Diseases* 23, 693–697.

16. Fain, A., and Smiley, R.L. (1989). A new cloacarid mite (Acari: Cloacaridae) from the lungs of the Great Horned Owl, *Bubo virginianus*, from the USA. *International Journal of Acarology* 15, 111–115.

17. Perez, T.M. (1996). The eggs of seven species of *Fainalges* Gaud and Berla (Xolalgidae) from the green conure (Aves: Psittacidae). In *Acarology IX*. Mitchell, R., Horn, D.J., Needham, G.R., and Welbourn, W.C. (Eds), 297–300. Columbus, OH: Ohio Biological Survey.

18. Mironov, S.V., and Proctor, H.C. (2008). The probable association of feather mites of the genus *Ingrassia* (Analgoidea: Xolalgidae) with the blue penguin *Eudyptula minor* (Aves: Sphenisciformes) in Australia. *Journal of Parasitology* 94, 1243–1248.

19. Gaud, J., and Atyeo, W.T. (1996). Feather mites of the world (Acarina, Astigmata): The supraspecific taxa. Part I: Text. *Annalen Zoologische Wetenschappen Koninklijk Museum voor Midden Afrika* 277, 1–193.

20. Choe, J.C., and Kim, K.C. (1989). Microhabitat selection and coexistence in feather mites (Acari: Analgoidea) on Alaskan seabirds. *Oecologia* 79, 10–14. Also: Choe, J.C., and Kim, K.C. (1991). Microhabitat selection of feather mites (Acari: Analgoidea) on murres and kittiwakes. *Canadian Journal of Zoology* 69, 817–821.

21. Wiles, P.R., Benke, C.J., Hartley, I.R., Gilbert, F.S., and McGregor, P.K. (2000). Season and ambient air temperature influence the distribution of mites (*Proctophylloides stylifer*) across the wings of Blue Tits (*Parus caeruleus*). *Canadian Journal of Zoology* 78, 1397–1407.

22. Jovani, R., and Serrano, D. (2001). Feather mites (Astigmata) avoid moulting wing feathers of passerine birds. *Animal Behaviour* 62, 723–727.

23. Jovani, R., Serrano, D., Frias, O., and Blanco, G. (2006). Molt imposes a trade-off on feather mite wing distribution: The case of barn swallows (*Hirundo rustica*). *Canadian Journal of Zoology* 84, 729–735.

24. Chen, B.L., Haith, K.L., and Mullens, B.A. (2011). Beak condition drives abundance and grooming-mediated competitive asymmetry in a poultry ectoparasite community. *Parasitology* 138, 748–757.

25. Proctor, H.C., and Owens, I. (2000). Mites and birds: Diversity, parasitism and coevolution. *Trends in Ecology and Evolution* 15, 358–364.

26. Galvan, I., and Sanz, J.J. (2006). Feather mite abundance increases with uropygial gland size and plumage yellowness in Great Tits, *Parus major*. *Ibis* 148, 687–697.

27. Bueter, C. (2005). Feather mites: A slight negative correlation with physiological condition. *Eukaryon* 1, 81–84.

28. Blanco, G., and Frias, O. (2001). Symbiotic feather mites synchronize dispersal and population growth with host sociality and migratory disposition. *Ecography* 24, 113–120.

29. Dowling, D.K., Richardson, D.S., and Komdeur, J. (2001). No effects of a feather mite on body condition, survivorship, or grooming behavior in the Seychelles warbler, *Acrocephalus sechellensis*. *Behavioral Ecology and Sociobiology* 50, 257–262.

30. Blanco, G., Tella, J.L., and Potti, J. (1997). Feather mites on group-living red-billed choughs: A non-parasitic interaction? *Journal of Avian Biology* 28, 197-206.

31. Brown, C.R., Brazeal, K.R., Strickler, S.A., and Brown, M.B. (2006). Feather mites are positively associated with daily survival in cliff swallows. *Canadian Journal of Zoology* 84, 1307-1314.

32. Atyeo, W.T., and Gaud, J. (1979). Feather mites and their hosts. In *Recent advances in acarology*. Rodriguez, J.G. (Ed), 355-361. New York: Academic Press.

33. DeVaney, J.A., and Augustine, P.C. (1988). Correlation of estimated and actual northern fowl mite populations with the evolution of specific antibody to a low molecular weight polypeptide in sera of infested hens. *Poultry Science* 67, 549-556.

34. Dabert, J., and Ehrnsberger, R. (1998). Phylogeny of the feather mite family Ptiloxenidae Gaud, 1982 (Acari: Pterolichoidea). *Biosystematics and Ecology Series* 14, 145-178.

35. Dyke, G.J., and van Tuinen, M. (2004). The evolutionary radiation of modern birds (Neornithes): Reconciling molecules, morphology and the fossil record. *Zoological Journal of the Linnean Society* 141, 153-177. *Also:* Hackett, S.J., Kimball, R.T., Reddy, S., Bowie, R.C.K., Braun, E.L., Braun, M.J., Chojnowski, J.L., Cox, W.A., Han, K.-L., Harshman, J., Huddleston, C.J., Marks, B.D., Miglia, K.J., Moore, W.S., Sheldon, F.H., Steadman, D.W., Witt, C.C., and Yuri, T. (2008). A phylogenetic study of birds reveals their evolutionary history. *Science* 320, 1763-1768.

36. Clayton, D.H., Bush, S.E., and Johnson, K.P. (2016). *Coevolution.of life of hosts: Integrating ecology and history*. Chicago: University of Chicago Press.

37. Proctor, H.C. (2001). *Megninia casuaricola* sp. n. (Acari: Analgidae), the first feather mite from a cassowary (Aves: Struthioniformes: Casuariidae). *Australian Journal of Entomology* 40, 335-341. *Also:* Mironov, S.V., and Proctor, H.C. (2005). A new feather mite genus of the family Psoroptoididae (Acari: Analgoidea) from cassowaries. *Journal of Natural History* 39, 3297-3304.

38. Mitchell, K., Llamas, B., Soubrier, J., Rawlence, N., Worthy, T., Wood, J., Lee, M., and Cooper, A. (2014). Ancient DNA reveals elephant birds and kiwi are sister taxa and clarifies ratite bird evolution. *Science* 344, 898-900.

39. Harlid, A., and Arnason, U. (1999). Analyses of mitochondrial DNA nest ratite birds within the Neognathae: Supporting a neotenous origin of ratite morphological characters. *Proceedings of the Royal Society of London. Series B: Biological Sciences* 266, 305-309.

40. DeBeer, G. (1956). The evolution of ratites. *Bulletin of the British Museum of Natural History (Zoology)* 4, 5-70.

41. Mironov, S.V., and Proctor, H.C. (2005). A new feather mite genus of the family Psoroptoididae (Acari: Analgoidea) from cassowaries. *Journal of Natural History* 39, 3297-3304.

6   FLYING ZOO FLIES

1. van Ripper, C., van Ripper, S.G., Goff, M.L., and Laird, M. (1986). The epizootiology and ecological significance of malaria in Hawaiian land birds. *Ecological Monographs* 56, 327-344.

2. Asghar, M., Hasselquist, D., Hansson, B., Zehtundjiev, P., Westerdahl, H., and Bensch, S. (2015). Hidden costs of infection: Chronic malaria accelerates telomere degradation and senescence in wild birds. *Science* 347, 436-438.

3. van Ripper, C., van Ripper, S.G., Goff, M.L., and Laird, M. (1986). The epizootiology and ecological significance of malaria in Hawaiian land birds. *Ecological Monographs* 56, 327-344.

4. Warner, R.E. (1968). The role of introduced diseases in the extinction of the endemic Hawaiian avifauna. *Condor* 70, 101-120.

5. Rohner, C., Krebs, C.J., Hunter, D.B., and Currie, D.C. (2000). Roost site selection of great horned owls in relation to black fly activity: An anti-parasite behavior? *Condor* 102, 950-955.

6. Hunter, D.B., Rohner, C., and Currie, D.C. (1997). Mortality in fledgling great horned owls from black fly hematophaga and leucocytozoonosis. *Journal of Wildlife Disease* 33, 486-491.

7. Sabrosky, C.W., Bennett, G.F., and Whitworth, T.L. (1989). *Bird blow flies (Protocalliphora) in North America (Diptera: Calliphoridae)*. Washington, DC: Smithsonian Institution Press.

8. Dawson, R.D., Hillen, K.K., and Whitworth, T.L. (2005). Effects of experimental variation in temperature on larval densities of parasitic *Protocalliphora* (Diptera: Calliphoridae) in nests of tree swallows (Passeriformes: Hirundinidae). *Environmental Entomology* 34, 563-568.

9. Whitworth, T.L., and Bennett, G.F. (1992). Pathogenicity of larval *Protocalliphora* (Diptera: Calliphoridae) parasitizing nestling birds. *Canadian Journal of Zoology* 70, 2184-2191.

10. Gold, C.S., and Dahlsten, D.L. (1983). Effects of parasitic flies (*Protocalliphora* spp.) on nestlings of mountain and chestnut-backed chickadees. *Wilson Bulletin* 95, 560-572.

11. Christe, P., Moller, A.P., and de Lope, F. (1998). Immunocompetence and nestling survival in the house martin: The tasty chick hypothesis. *Oikos* 83, 175-179.

12. Simon, A., Thomas, D.W., Blondel, J., Perret, P., and Lambrechts, M.M. (2003). Within-brood distribution of ectoparasite attacks on nestling blue tits: A test of the tasty chick hypothesis using insulin as a tracer. *Oikos* 102, 551-558.

13. Decamps, S., Blondel, J., Lambrechts, M.M., Hurtez-Bousses, S., and Thomas, F. (2002). Asynchronous hatching in a blue tit population: A test of some predictions related to ectoparasites. *Canadian Journal of Zoology* 80, 1480-1484.

14. Roulin, A., Brinkhof, M.W.G., Bize, P., Richner, H., Jungi, T.W., Bavoux, C., Boileau, N., and Burneleau, G. (2003). Which chick is tasty to parasites? The importance of host immunology vs. parasite life history. *Journal of Animal Ecology* 72, 75-81.

15. Valera, F., Hoi, H., Darolova, A., and Kristofik, J. (2004). Size versus health as a cue for host choice: A test of the tasty chick hypothesis. *Parasitology* 129, 59-68.

16. Hurtrez-Bousses, S., Blondel, J., Perret, P., Fabreguettes, J., and Renaud, F. (1998). Chick parasitism by blowflies affects feeding rates in a Mediterranean population of blue tits. *Ecological Letters* 1, 17-20.

17. Spalding, M.G., Mertens, J.W., Walsh, P.B., Morin, K.C., Dunmore, D.E., and Forrester, D.J. (2002). Burrowing fly larvae (*Philornis porteri*) associated with mortality of eastern bluebirds in Florida. *Journal of Wildlife Disease* 38, 776-783.

18. Smith, N.G. (1968). The advantage of being parasitized. *Nature* 219, 690–694.

19. Hilton, B. Jr. (2005). Stealth flies on birds: The hippoboscids. *This week at Hilton Pond.* http://www.hiltonpond.org/ThisWeek051108.html.

20. Rothschild, M., and Clay, T. (1952). *Fleas, flukes and cuckoos: A study of bird parasites*, p. 212. London, UK: Philosophical Library.

21. Jovani, R., Tella, J.L., Sol, D., and Ventura, D. (2001). Are hippoboscid flies a major mode of transmission of feather mites? *Journal of Parasitology* 87, 1187–1189.

22. Whiteman, N.K., Sanchez, P., Merkel, J., Klompen, H., and Parker, P.G. (2006). Cryptic host specificity of an avian skin mite (Epidermoptidae) vectored by louseflies (Hippoboscidae) associated with two endemic Galapagos bird species. *Journal of Parasitology* 92, 1218–1228.

23. Hodgson, J.C., Spielman, A., Komar, N., Krahforst, C.F., Wallace, G.T., and Pollack, R.J. (2001). Interrupted blood-feeding by *Culiseta melanura* (Diptera: Culicidae) on European starlings. *Journal of Medical Entomology* 38, 59–66.

24. Douglas, H.D., Co, J.E., Jones, T.H., Connor, W.E., and Day, J.F. (2005). Chemical odorant of colonial seabird repels mosquitoes. *Journal of Medical Entomology* 42, 647–651.

25. Pierotti, R. (1991). Infanticide versus adoption: An intergenerational conflict. *American Naturalist* 138, 1140–1158.

26. Bize, P., Roulin, A., and Richner, H. (2003). Adoption as an offspring strategy to reduce ectoparasite exposure. *Proceedings of the Royal Society of London. Series B: Biological Sciences* 270, S114–S116.

27. Borkent, A., Coram, R.A., and Jarzembowski, E.A. (2013). The oldest fossil biting midge (Diptera: Ceratopogonidae) from the Purbeck Limestone Group (Lower Cretaceous) of southern Great Britain. *Polish Journal of Entomology* 82, 273–279.

28. Poinar, G. Jr. (2016). What fossils reveal about the protozoa progenitors, geographic provinces, and early hosts of malarial organisms. *American Entomologist* 62, 22–25.

7   THE WORMS THAT ATE THE BIRD

1. Bush, A.O., Aho, J.M., and Kennedy, C.R. (1990). Ecological versus phylogenetic determinants of helminth parasite community richness. *Evolutionary Ecology* 4, 1–20. *Also*: Kennedy, C.R., Bush, A.O., and Aho, J.M. (1986). Patterns in helminth communities: Why are birds and fish different? *Parasitology* 93, 205–215. *Also*: Poulin, R. (2007). *Evolutionary ecology of parasites, 2nd ed.* Princeton, NJ: Princeton University Press.

2. Poulin, R. (2007). *Evolutionary ecology of parasites, 2nd ed.* Princeton, NJ: Princeton University Press.

3. Bush, A.O., and Kennedy, C.R. (1994). Host fragmentation and helminth parasites: Hedging your bets against extinction. *International Journal of Parasitology* 24, 1333–1343. *Also*: Poulin, R. (1998). Large scale patterns of host use by parasites of freshwater fishes. *Ecological Letters* 1, 118–128.

4. Bush, A.O., and Kennedy, C.R. (1994). Host fragmentation and helminth parasites: Hedging your bets against extinction. *International Journal of Parasitology* 24, 1333–1343.

5. Bush, A.O., and Lotz, J.M. (2000). The ecology of "crowding." *Journal of Parasitology* 86, 212–213.

6. Bush, A.O., and Holmes, J.C. (1986). Intestinal helminthes of lesser scaup ducks: An interactive community. *Canadian Journal of Zoology* 64, 142–152. *Also*: Butterworth, E.W. (1982). *A study of the structure and organization of intestinal helminth communities in ten species of waterfowl (Anatinae)*. PHD dissertation, University of Alberta.

7. Stock, T.M., and Holmes, J.C. (1987). *Dioecocestus asper* (Cestoda: Dioecocestidae): An interference competitor in an enteric helminth community. *Journal of Parasitology* 73, 1116–1123. *Also*: Stock, T.M., and Holmes, J.C. (1988). Functional relationships and microhabitat distributions of enteric helminthes of grebes (Podicipedidae): The evidence for interactive communities. *Journal of Parasitology* 74, 214–227.

8. Bethel, W.M., and Holmes, J.C. (1973). Altered evasive behavior and responses to light in amphipods harboring acanthocephalan cystacanths. *Journal of Parasitology* 39, 945–956.

9. Helluy, S.M., and Holmes, J.C. (1990). Serotonin, octopamine, and the clinging behavior induced by the parasite *Polymorphus paradoxus* (Acanthocephala) in *Gammarus lacustris* (Crustacea). *Canadian Journal of Zoology* 68, 1214–1220.

10. Maynard, B.J., DeMartini, L., and Wright, W.G. (1996). *Gammarus lacustris* harboring *Polymorphus paradoxus* show altered patterns of serotonin-like immunoreactivity. *Journal of Parasitology* 82, 663–666.

11. For a review of this research, see: Moore, J., and Gotelli, N.J. (1990). Phylogenetic perspective on the evolution of altered host behaviours: A critical look at the manipulation hypothesis. In *Parasitism and host behaviour*. Barnard, C.J., and Behnke, J.M. (Eds), 193–233. London, UK: Taylor and Francis.

12. Holmes, J.C., and Bethel, W.M. (1972). Modification of intermediate host behaviour by parasites. In *Behavioural aspects of parasite transmission*. Canning, E.U., and Wright, C.A. (Eds), 123–149. London, UK: Academic Press.

13. Barber, I., Walker, P., and Svensson, P.A. (2004). Behavioural responses to simulated avian predation in female three-spined sticklebacks: The effect of experimental *Schistocephalus solidus* infections. *Behaviour* 141, 1425–1440. *Also*: Loot, G., Aulagnier, S., Lek, S., Thomas, F., and Guegan, J.-F. (2002). Experimental demonstration of a behavioural modification in a cyprinid fish, *Rutilus rutilus* (L.), induced by a parasite, *Ligula intestinalis* (L.). *Canadian Journal of Zoology* 80, 738–744.

14. Loot, G., Lek, S., Brown, S.P., and Guegan, J.-F. (2001). Phenotypic modification of roach (*Rutilus rutilus* L.) infected with *Ligula intestinalis* L. (Cestoda: Pseudophyllidea). *Journal of Parasitology* 87, 1002–1010.

15. Van Dobben, W.H. (1952). The food of the cormorant in the Netherlands. *Ardea* 40, 1–63.

16. Holmes, J.C., and Bethel, W.M. (1972). Modification of intermediate host behaviour by parasites. In *Behavioural aspects of parasite transmission*. Canning, E.U., and Wright, C.A. (Eds), 123–149. London, UK: Academic Press.

17. Bush, A.O., and Holmes, J.C. (1986). Intestinal helminthes of lesser scaup ducks: Patterns of associations. *Canadian Journal of Zoology* 64, 132–141. *Also*: Stock, T.M., and Holmes, J.C. (1988). Functional relationships and microhabitat distributions of enteric helminthes of grebes (Podicipedidae): The evidence for interactive communities. *Journal of Parasitology* 74, 214–227.

18. Butterworth, E.W. (1982). *A study of the structure and organization of intestinal helminth communities in ten species of waterfowl (Anatinae)*. PHD Dissertation, University of Alberta.

19. Bush, A.O., and Holmes, J.C. (1986). Intestinal helminthes of lesser scaup ducks: Patterns of associations. *Canadian Journal of Zoology* 64, 132–141.

20. Stock, T.M., and Holmes, J.C. (1987). *Dioecocestus asper* (Cestoda: Dioecocestidae): An interference competitor in an enteric helminth community. *Journal of Parasitology* 73, 1116–1123.

21. Cracraft, J. (1982). Phylogenetic relationships and monophyly of loons, grebes and hesperornithiform birds, with comments on the early history of birds. *Systematic Zoology* 31, 35–56.

22. Czaplinski, B. (1975). Hymenolepididae parasitizing mute swans *Cygnus olor* (Gm.) of different age, in Poland. *Acta Parasitologica Polonica* 23, 305–327. *Also:* Kotecki, N.R. (1970). Circulation of the cestode fauna of Anseriformes in the Municipal Zoological Garden of Warsaw. *Acta Parasitologica Polonica* 17, 329–355.

23. Mitter, C., and Brooks, D.R. (1983). Phylogenetic aspects of coevolution. In *Coevolution*. Futuyma, D.J., and Slatkin, M. (Eds), 65–98. Sunderland, MA: Sinauer.

24. Stock, T.M., and Holmes, J.C. (1988). Functional relationships and microhabitat distributions of enteric helminthes of grebes (Podicipedidae): The evidence for interactive communities. *Journal of Parasitology* 74, 214–227.

25. Paszkowski, C.A., Gingras, B.A., Wilcox, K., Klatt, P.H., and Tonn, W.M. (2004). Trophic relations of the red-necked grebe on lakes in the western boreal forest: A stable-isotope analysis. *Condor* 106, 638–651. *Also:* McParland, C.E., Paszkowski, C.A., and Newbrey, J.L. (2010). Trophic relationships of breeding Red-necked grebes (Podiceps grisegena) on wetlands with and without fish in the Aspen Parkland. *Canadian Journal of Zoology* 88, 186–194.

26. Korpaczewska, W., and Sulgostowska, T. (1974). Revision of the genus *Tatria* Kow., 1904 (Cestoda: Amabiliidae), including a description of *Tatria inuii* sp. n. *Acta Parasitologica Polonica* 22, 67–91.

27. Sibley, C.G., and Ahlquist, J.E. (1990). *Phylogeny and classification of birds: A study in molecular evolution.* New Haven, CT: Yale University Press.

28. Olsen, S.L., and Feduccia, A. (1980). Relationships and evolution of flamingos (Aves: Phoenicopteridae). *Smithsonian Contributions to Zoology* 316, 1–73.

29. Van Tuinen, M., Butvill, D.B., Kirsch, J.A.W., and Hedges, S.B. (2001). Convergence and divergence in the evolution of aquatic birds. *Proceedings of the Royal Society of London. Series B: Biological Sciences* 268, 1345–1350. *Also:* Poe, S., and Chubb, A.L. (2004). Birds in a bush: Five genes indicate explosive evolution of avian orders. *Evolution* 58, 404–415. *Also:* Mayr, G. (2004). Morphological evidence for sister group relationship between flamingos (Aves: Phoenicopteridae) and grebes (Podicipedidae). *Zoological Journal of the Linnean Society* 140, 157–169.

30. Sangster, G. (2005). A name for the flamingo-grebe clade. *Ibis* 147, 612–615.

31. Storer, R.W. (2006). The grebe-flamingo connection: A rebuttal. *The Auk* 123, 1183–1184.

32. Hopkins, G.H.E. (1942). The mallophaga as an aid to the classification of birds. *Ibis* 6, 94–106.

33. Johnson, K.P., Kennedy, M., and McCracken, K.G. (2006). Reinterpreting the origins of flamingo lice: Cospeciation or host-switching? *Biological Letters* 2, 275–278.

34. Rudolfova, J., Littlewood, D.T.J., Sitko, J., and Horak, P. (2007). Bird schistosomes of wildfowl in the Czech Republic and Poland. *Folia Parasitologica* 54, 88–93.

1. Hilgartner, R., Raoilison, M., Buttiker, W., Lees, D.C., and Krenn, H.W. (2007). Malagasy birds as hosts for eye-frequenting moths. *Biological Letters* 3, 117–120.

2. Riley, J. (1986). The biology of Pentastomids. *Advances in Parasitology* 25, 46–128. *Also:* Ruppert, E.E., Fox, R.S., and Barnes, R.D. (2004). *Invertebrate zoology: A functional evolutionary approach.* New York: Thomson Brooks/Cole.

3. Riley, J., Oaks, J.L., and Gilbert, M. (2003). *Raillietiella trachea* n. sp., a pentastomid from the trachea of an oriental white-backed vulture *Gyps bengalensis* taken in Pakistan, with speculation about its life-cycle. *Systematic Parasitology* 56, 155–161. *Also:* Martinez, J., Criado-Fornelio, A., Lanzarot, P., Fernandez-Garcia, M., Rodriguez-Caabeiro, F., and Merino, S. (2004). A new pentastomid from the black vulture. *Journal of Parasitology* 90, 1103–1105.

4. Bakke, T.A. (1972). *Reighardia sternae* (Diesing, 1864) Ward 1899 (Pentastomida: Cephalobaenida) from the common gull (*Larus canus* L.) in a Norwegian locality. *Norwegian Journal of Zoology* 20, 273–277.

5. Banaja, A.A., James, J.L., and Riley, J. (1975). An experimental investigation of a direct life-cycle in *Reighardia sternae* (Diesing, 1864), a pentastomid parasite of the herring gull (*Larus argentatus*). *Parasitology* 71, 493–503.

6. Waldbaurer, G. (1998). *The birder's bug book.* Cambridge, MA: Harvard University Press.

7. Grimaldi, D.A., Engel, M.S., and Nascimbene, P.C. (2002). Fossiliferous Cretaceous amber from Myanmar (Burma): Its rediscovery, biotic diversity, and paleontological significance. *American Museum Novitates* 3361, 1–71.

8. Poinar, G. Jr. (2016). What fossils reveal about the protozoa progenitors, geographic provinces, and early hosts of malarial organisms. *American Entomologist* 62, 22–25. *Also:* Poinar, G. Jr., and Poinar, R. (2007). *What bugged the dinosaurs? Insects, disease, and death in the Cretaceous.* Princeton, NJ: Princeton University Press.

9. Brown, C.R., and Brown, M.B. (1986). Ectoparasitism as a cost of coloniality in cliff swallows (*Hirundo pyrrhonota*). *Ecology* 67, 1206–1218.

10. Chapman, B.R., and George, J.E. (1991). The effects of ectoparasites on cliff swallow growth and survival. In *Bird-parasite interactions: Ecology, evolution and behaviour.* Loye, J.E., and Zuk, M. (Eds), 69–92. Oxford: Oxford University Press.

11. Loye, J.E., and Carroll, S.P. (1991). Nest ectoparasite abundance and cliff swallow colony site selection, nestling development, and departure time. In *Bird-parasite interactions: Ecology, evolution and behaviour.* Loye, J.E., and Zuk, M. (Eds), 222–241. Oxford: Oxford University Press.

12. Chapman, B.R., and George, J.E. (1991). The effects of ectoparasites on cliff swallow growth and survival. In *Bird-parasite interactions: Ecology, evolution and behaviour.* Loye, J.E., and Zuk, M. (Eds), 69–92. Oxford: Oxford University Press.

13. Brown, C.R., and Brown, M.B. (2002). Ectoparasites cause increased bilateral asymmetry of naturally selected traits in a colonial bird. *Journal of Evolutionary Biology* 15, 1067–1075.

14. Loye, J.E., and Carroll, S.P. (1991). Nest ectoparasite abundance and cliff swallow colony site selection, nestling development, and departure time. In *Bird-parasite interactions: Ecology, evolution and behaviour.* Loye, J.E., and Zuk, M. (Eds), 222–241. Oxford: Oxford University Press.

15. Trauger, D.L., and Bartonek, J.C. (1977). Leech parasitism of waterfowl in North America. *Wildfowl* 28, 143–152.
16. Tuggle, B.N. (1986). The occurrence of Theromyzon rude (Annelida: Hirudinea) in association with mortality of trumpeter swan cygnets (Cygnus buccinator). *Journal of Wildlife Disease* 22, 279–280.
17. Miller, C. (1997). Occurrence and ecology of the Open Bay Islands leech, *Hirudobdella antipodium*. *Science for Conservation* 57, 1–16.
18. Hamilton, P.B., Stevens, J.R., Gidley, J., Holz, P., and Gibson, W.C. (2005). A new lineage of trypanosomes from Australian vertebrates and terrestrial bloodsucking leeches (Haemadipsidae). *International Journal of Parasitology* 35, 431–443.
19. Stirling, I., and Johns, P.M. (1969). Notes on the bird fauna of Open Bay Islands. *Notornis* 16, 121–125.
20. Miller, C. (1997). Occurrence and ecology of the Open Bay Islands leech, *Hirudobdella antipodium*. *Science for Conservation* 57, 1–16.
21. Kumara, T., Masayuki, H., Takahisa, Y., and Tatsumi, N. (2000). Hirudin as an anticoagulant for both haematology and chemistry tests. *Journal of Automated Methods and Management in Chemistry* 22, 109–112.

9   FLYING ZOO BEHAVIOUR

1. Anderson, R.M., and May, R.M. (1979). Population biology of infectious diseases: Part I. *Nature* 280, 361–367.
2. Hart, B.L. (1990). Behavioral adaptations to pathogens and parasites: Five strategies. *Neuroscience and Biobehavioral Reviews* 14, 273–294.
3. Clark, L., and Mason, J.R. (1988). Effect of biologically active plants used as nest material and the derived benefit to starling nestlings. *Oecologia* 77, 174–180.
4. Brown, N.S. (1972). The effect of host beak condition on the size of *Menacanthus stramineus* population of domestic chickens. *Poultry Science* 51, 162–164.
5. Brooke, M. de L. (1985). The effect of allopreening on tick burdens of molting eudyptid penguins. *The Auk* 102, 893–895.
6. Hamilton, W.D., and Zuk, M. (1982). Heritable true fitness and bright birds: A role for parasites? *Science* 218, 384–387.
7. Read, A.F. (1988). Sexual selection and the role of parasites. *Trends in Ecology and Evolution* 3, 97–102.
8. Zuk, M. (1992). The role of parasites in sexual selection: Current evidence and future directions. *Advances in the Study of Behavior* 21, 39–68.
9. Moller, A.P. (1991). Parasites, sexual ornaments and mate choice in the barn swallow *Hirundo rustica*. In *Bird-parasite interactions: Ecology, evolution, and behaviour*. Loye, J.E., and Zuk, M. (Eds), 328–343. Oxford: Oxford University Press.
10. Moller, A.P. (1988). Female choice selects for male sexual tail ornaments in the monogamous swallow. *Nature* 332, 640–642.
11. Johnson, L.L., and Boyce, M.S. (1991). Female choice of males with low parasite loads in sage grouse. In *Bird-parasite interactions: Ecology, evolution, and behaviour*. Loye, J.E., and Zuk, M. (Eds), 377–388. Oxford: Oxford University Press.

12. Boyce, M.S. (1990). The Red Queen visits sage grouse leks. *American Zoologist* 30, 266-278.

13. Spurrier, M.F., Boyce, M.S., and Manly, B.F.J. (1991). Effects of parasites on mate choice by captive sage grouse. In *Bird-parasite interactions: Ecology, evolution, and behaviour.* Loye, J.E., and Zuk, M. (Eds), 389-398. Oxford: Oxford University Press.

14. Catchpole, C.K., and Slater, P.J.B. (1995). *Bird song: Biological themes and variations.* Cambridge: Cambridge University Press.

15. Nowicki, S., Peters, S., and Podos, J. (1998). Song learning, early nutrition and sexual selection in songbirds. *American Zoologist* 38, 179-190. *Also:* Nowicki, S., Searcy, W.A., and Peters, A. (2002). Brain development, song learning and mate choice in birds: A review and experimental test of the nutritional stress hypothesis *Journal of Comparative Physiology A: Neuroethology, Sensory, Neural, and Behavioral Physiology* 188, 1003-1014. *Also:* Devoogd, T.J., Krebs, J.R., Healy, S.D., and Purvis, A. (1993). Relations between song repertoire size and volume of brain nuclei related to song: Comparative evolutionary analyses amongst oscine birds. *Proceedings of the Royal Society of London. Series B: Biological Sciences* 254, 75-82.

16. Spencer, K.A., Buchanan, K.L., Leitner, S., Goldsmith, A.R., and Catchpole, C.K. (2005). Parasites affect song complexity and neural development in a songbird. *Proceedings of the Royal Society of London. Series B: Biological Sciences* 272, 2037-2043.

17. Buchanan, K.L., Catchpole, C.K., Lewis, J.W., and Lodge, A. (1999). Song as an indicator of parasitism in the sedge warbler. *Animal Behaviour* 57, 307-314.

18. Hasselquist, D., Bensch, S., and von Schantz, T. (1996). Correlation between song repertoire, extra-pair paternity and offspring survival in the great reed warbler. *Nature* 381, 229-232.

19. Webster, L.M.I., Johnson, P.C.D., Adam, A., Mable, B.K., and Keller, L.K. (2007). Macrogeographic population structure in a parasitic nematode with avian hosts. *Veterinary Parasitology* 144, 93-103.

20. Tompkins, D.M., and Hudson, P.J. (2000). Field evidence for apparent competition mediated via the shared parasites of two gamebird species. *Ecological Letters* 3, 10-14.

21. Mougeot, F., Redpath, S.M., and Leckie, F. (2005). Ultra-violet reflectance of male and female red grouse, *Lagopus lagopus scoticus:* Sexual ornaments reflect nematode parasite intensity. *Journal of Avian Biology* 36, 203-209.

22. Mougeot, F., and Redpath, S.M. (2004). Sexual ornamentation relates to immune function in male red grouse *Lagopus lagopus scoticus. Journal of Avian Biology* 35, 425-433.

23. Fox, A., and Hudson, P.J. (2001). Parasites reduce territorial behaviour in red grouse (*Lagopus lagopus scoticus*). *Ecological Letters* 4, 139-143.

24. Galvan I., and Sanz, J.J. (2006). Feather mite abundance increases with uropygial gland size and plumage yellowness in Great Tits *Parus major. Ibis* 148, 687-697.

25. Harper, D.G.C. (1999). Feather mites, pectoral muscle condition, wing length and plumage coloration of passerines. *Animal Behaviour* 58, 553-562.

26. John, J.L. (1997). The Hamilton-Zuk theory and initial test: An examination of some parasitological criticisms. *International Journal of Parasitology* 27, 1269-1288.

27. Hill, G.E. (1999). Is there an immunological cost to carotenoid-based ornamental coloration? *American Naturalist* 154, 589-595. *Also:* Hill, G.E. (1999). Mate choice,

male quality, and carotenoid-based plumage coloration. *Proceedings of the International Ornithological Congress* 22, 1–11.

28. Lozano, G. (1994). Carotenoids, parasites and sexual selection. *Oikos* 70, 309–311.

29. Saino, N., Ferrari, R., Romano, M., Martinelli, R., and Moller, A.P. (2003). Experimental manipulation of egg carotenoids affects immunity of barn swallow nestlings. *Proceedings of the Royal Society of London. Series B: Biological Sciences* 270, 2485–2489.

30. Navara, K.J., Badyaev, A.V., Mendonca, M.T., and Hill, G.E. (2006). Yolk antioxidants vary with male attractiveness and female condition in the house finch (*Carpodacus mexicanus*). *Physiological and Biochemical Zoology* 79, 1098–1105.

31. Norris, K., and Evans, M.R. (2000). Ecological immunology: Life history trade-offs and immune defense in birds. *Behavioral Ecology* 11, 19–26.

32. Saino, N., Calza, S., and Moller, A.P. (1997). Immunocompetence of nestling barn swallows (*Hirundo rustica*) in relation to brood size and parental effort. *Journal of Animal Ecology* 66, 827–836.

33. Verhulst, S., Dieleman, S.J., and Parmentier, H.K. (1999). A trade-off between immunocompetence and sexual ornamentation in domestic fowl. *Proceedings of the National Academy of Sciences* 96, 4478–4481.

34. McGraw, K.J. (2006). Mechanics of uncommon colors: Pterins, poprphyrins and psittacofulvins. In *Bird coloration, vol. 1: Mechanics and measurements*. Hill, G.E., and McGraw, K.J. (Eds), 354–398. Cambridge, MA: Harvard University Press.

35. Masello, J.F., and Quillfeldt, P. (2003). Body size, body condition and ornamental feathers of burrowing parrots: Variation between years and sexes, assortative mating and influences on breeding success. *Emu* 103, 149–161.

36. Veiga, J.P., and Puerta, M. (1996). Nutritional constraints determine the expression of a sexual trait in the House Sparrow, *Passer domesticus. Proceedings of the Royal Society of London. Series B: Biological Sciences* 263, 229–234.

37. Fitze, P.S., and Richner, H. (2002). Differential effects of a parasite on ornamental structures based on melanins and carotenoids. *Behavioral Ecology* 13, 401–407.

38. Jawor, J.M., and Breitwisch, R. (2003). Melanin ornaments, honesty, and sexual selection. *The Auk* 120, 249–265.

39. Doucet, S., and Montgomerie, R. (2003). Structural plumage and parasites in satin bowerbirds: Implications for sexual selection. *Journal of Avian Biology* 34, 237–242.

40. Griffith, S.C., and Pryke, S.R. (2006). Benefits to females assessing color displays. In *Bird coloration, vol. 2: Function and evolution*. Hill, G.E., and McGraw, K.J. (Eds), 233–279. Cambridge, MA: Harvard University Press.

41. Chakra, M.A., Hilbe, C., and Traulsen, A. (2014). Plastic behaviors in hosts promote the emergence of retaliatory parasites. *Science Reports* 4, 4251.

42. Gluckman, T.L., and Mundy, N.I. (2013). Cuckoos in raptors' clothing: Barred plumage illustrates a fundamental principle of Batesian mimicry. *Animal Behaviour* 86, 1165–1181.

43. Feeney, W.E., Troscianko, J., Langmore, N.E., and Spottiswoode, C.N. (2015). Evidence for aggressive mimicry in an adult brood parasitic bird, and generalized defences in its host. *Proceedings of the Royal Society of London. Series B: Biological Sciences* 282, 20150795.

44. Caves, E.M., Stevens, M., Iversen, E., and Spottiswoode, C.N. (2015). Hosts of avian brood parasites have evolved egg signatures with elevated information content. *Proceedings of the Royal Society of London. Series B: Biological Sciences* 282, 20150598.

45. Tokue, K., and Ueda, K. (2010). Mangrove gerygones *Gerygone laevigaster* eject little bronze-cuckoo *Chalcites minutillus* hatchlings from parasitized nests. *Ibis* 152, 835–839.

46. Colombelli-Negrel, D., Hauber, M.E., Robertson, J., Sulloway, F.J., Hoi, H., Griggio, M., and Kleindorfer, S. (2012). Embryonic learning of vocal passwords in superb fairy-wrens reveals intruder cuckoo nestlings. *Current Biology* 22, 2155–2160.

47. Thorogood, R., and Davies, N.B. (2016). Combining personal with social information facilitates host defences and explains why cuckoos should be secretive. *Science Reports* 6, 19872.

48. Sherry, D.F., Forbes, M.R.L., Khurgel, M., and Ivy, G.O. (1993). Females have a larger hippocampus than males in the brood-parasitic brown-headed cowbird. *Proceedings of the National Academy of Sciences* 90, 7839–7843.

49. Canestrari, D., Bolopo, D., Turlings, T.C.J., Roder, G., Marcos, J.M., and Baglione, V. (2014). From parasitism to mutualism: Unexpected interactions between a cuckoo and its host. *Science* 343, 1350–1352.

10   ENVIRONMENTAL IMPACTS
The Future of the Flying Zoo

1. Price, P.W. (1980). *Evolutionary biology of parasites: Monographs in population biology, vol. 15.* Princeton, NJ: Princeton University Press.

2. Dobson, A., Lafferty, K.D., Kuris, A.M., Hechinger, R.F., and Jetz, W. (2008). Homage to Linnaeus: How many parasites? How many hosts? *Proceedings of the National Academy of Sciences* 105, 11482–11489.

3. Dobson, A., Lafferty, K.D., Kuris, A.M., Hechinger, R.F., and Jetz, W. (2008). Homage to Linnaeus: How many parasites? How many hosts? *Proceedings of the National Academy of Sciences* 105, 11482–11489.

4. Wilson, E.O. (1992). *The diversity of life.* Cambridge, MA: Belknap Press.

5. Marra, P.P., Griffing, S., Caffrey, C., Kilpatrick, A.M., McLean, R., Brand, C., Saito, E., Dupuis, A.P., Kramer, L., and Novak, R. (2004). West Nile virus and wildlife. *BioScience* 54, 393–402.

6. Colpitts, T.M., Conway, M.J., Montgomery, R.R., and Fikrig, E. (2012). West Nile virus: Biology, transmission and human infection. *Clinical Microbiology Review* 25, 635–648.

7. Dohm, D.J., O'Guinn, M.L., and Turell, M.J. (2002). Effect of environmental temperature on the ability of *Culex pipiens* (Diptera: Culicidae) to transmit West Nile virus. *Journal of Medical Entomology* 39, 221–225.

8. van Ripper, C. III, van Ripper, S.G., Goff, M.L., and Laird, M. (1986). The epizootiology and ecological significance of malaria in Hawaiian land birds. *Ecological Monographs* 56, 327–344.

9. Woodworth, B.L., Atkinson, C.T., LaPoint, D.A., Hart, P.J., Spiegel, C.S., Tweed, E.J., Henneman, C., LeBrun, J., Danette, T., DeMots, R., Kozar, K.L., Triglia, D., Lease, D., Gregor, A., Smith, T., and Duffy, D. (2005). Host population persistence in the face of introduced vector-borne diseases: Hawaii amakihi and avian malaria. *Proceedings of the National Academy of Sciences* 102, 1531–1536.

10. Atkinson, C.T., Saili, K.S., Utzurrum, R.B., and Jarvi, S.I. (2013). Experimental evidence for evolved tolerance to avian malaria in a wild population of low elevation Hawai'i 'Amakihi (*Hemignathus virens*). *EcoHealth* 10, 366-375.

11. Callaghan, T.V., Bjorn, L.O., Chernov, Y., Chapin, T., Christensen, T.R., Huntley, B., Ims, R.A., Johansson, M., Jolly, D., Jonasson, S., Matveyeva, N., Panikov, N., Oechel, W., Shaver, G., Elster, J., Henttonen, H., Laine, K., Taulavuori, K., Taulavuori, E., and Zockler, C. (2004). Biodiversity, distributions and adaptations of Arctic species in the context of environmental change. *Ambio* 33, 404-417.

12. Comiso, J.C., and Hall, D.K. (2014). Climate trends in the Arctic as observed from space. *Wiley Interdisciplinary Reviews: Climate Change* 5, 389-409.

13. Callaghan, T.V., Bjorn, L.O., Chernov, Y., Chapin, T., Christensen, T.R., Huntley, B., Ims, R.A., Johansson, M., Jolly, D., Jonasson, S., Matveyeva, N., Panikov, N., Oechel, W., Shaver, G., Elster, J., Henttonen, H., Laine, K., Taulavuori, K., Taulavuori, E., and Zockler, C. (2004). Biodiversity, distributions and adaptations of Arctic species in the context of environmental change. *Ambio* 33, 404-417.

14. Wrona, F.J., Prowse, T.D., Reist, J.D., Hobbie, J.E., Lévesque, L.M.J., and Vincent, W.F. (2006). Climate change effects on aquatic biota, ecosystem structure and function. *Ambio* 35, 359-369.

15. Samuel, W.M., Pybus, M.J., Welch, D.A., and Wilke, C.J. (1992). Elk as a potential host for meningeal worm: Implications for translocation. *Journal of Wildlife Management* 56, 629-639.

16. Fitter, J., Fitter, D., and Hosking, D. (2016). *Wildlife of the Galapagos, 2nd ed.* Princeton, NJ: Princeton University Press.

17. Gottdenker, N.L., Walsh, T., Vargas, H., Merkel, J., Jimenez, G.U., Miller, R.E., Diley, M., and Parker, P.G. (2005). Assessing the risks of introduced chickens and their pathogens to native birds in the Galapagos Archipelago. *Biological Conservation* 126, 429-439.

18. Docherty, D.E., and Friend, M. (1999). Newcastle disease. In *Field Manual of Wildlife Diseases: General Field Procedures and Diseases of Birds*. Friend, M., and Franson, J.C. (Eds), 175-179. Madison, WI: US Geological Survey-National Wildlife Health Center.

19. Whiteman, N.K., Goodman, S.J., Sinclair, B.J., Walsh, T., Cunningham, A.A., Kramer, L.D., and Parker, P.G. (2005). Establishment of the avian disease vector *Culex quinquefasciatus* Say, 1823 (Diptera: Culicidae) on the Galapagos Islands, Ecuador. *Ibis* 147, 844-847.

20. Wikelski, M., Foufopoulos, J., Vargas, H., and Snell, H. (2004). Galapagos birds and diseases: Invasive pathogens as threats for island species. *Ecology and Society* 9, 5-14.

21. Fessl, B., Kleindorfer, S., and Tebbich, S. (2005). An experimental study on the effects of an introduced parasite in Darwin's finches. *Biological Conservation* 127, 55-61.

22. Wikelski, M., Foufopoulos, J., Vargas, H., and Snell, H. (2004). Galapagos birds and diseases: Invasive pathogens as threats for island species. *Ecology and Society* 9, 5-14.

23. Knutie, S.A., McNew, S.M., Bartlow, A.W., Vargas, D.A., and Clayton, D.H. (2014). Darwin's finches combat introduced nest parasites with fumigated cotton. *Current Biology* 24, R355-R356.

24. Parker, P.G., Whiteman, N.K., and Miller, R.E. (2006). Conservation medicine on the Galapagos Islands: Partnerships among behavioral, population, and veterinary scientists. *The Auk* 123, 625-638.

25. Carson, R. (1962). *Silent spring*. Boston: Houghton Mifflin.

26. Sagerup, K., Henriksen, E.O., Skorping, A., Skaare, J.U., and Gabrielsen, G.W. (2000). Intensity of parasitic nematodes increases with organochlorine levels in the glaucous gull. *Journal of Applied Ecology* 37, 532–539.

27. Bustnes, J.O., Erikstad, K.E., Hanssen, S.A., Tveraa, T., Folstad, I., and Skaare, J.U. (2006). Anti-parasite treatment removes negative effects of environmental pollutants on reproduction in an Arctic seabird. *Proceedings of the Royal Society of London. Series B: Biological Sciences* 273, 3117–3122.

28. For a personal account, see: Stutchbury, B. (2007). *Silence of the songbirds: How we are losing the world's songbirds and what we can do to save them.* New York: HarperCollins.

29. Stutchbury, B. (2007). *Silence of the songbirds: How we are losing the world's songbirds and what we can do to save them.* New York: HarperCollins.

30. Kelly S.T., and DeCapita, M.E. (1982) Cowbird control and its effect on Kirtland's warbler reproductive success. *Wilson Bulletin* 94, 363–365.

31. Ostfeld, R.S., and Keesing, F. (2000). Biodiversity and disease risk: The case of Lyme disease. *Conservation Biology* 14, 722–728.

32. Alberta Government. (2015). *Surveillance of ticks on companion animals in Alberta: 2015 Summary.* http://www1.agric.gov.ab.ca/$Department/deptdocs.nsf/all/cpv13168/$FILE/2015-tick-summary.pdf.

33. Cizauskas, C.A., Carlson, C.J., Burgio, K.R., Clements, C.F., Dougherty, E.R., Harris, N.C., and Phillips, A.J. (2017). Parasite vulnerability to climate change: An evidence-based functional trait approach. *Royal Society Open Science* 4, 106535.

34. Pizzi, R. (2009). Veterinarians and taxonomic chauvinism: The dilemma of parasite conservation. *Journal of Exotic Pet Medicine* 18, 279–282.

35. Holmes, J.C. (1996). Parasites as threats to biodiversity in shrinking ecosystems. *Biodiversity and Conservation* 5, 975–983.

FURTHER

READING

1   A WORLD ON A BIRD

Bush, A.O., Fernandez, J.C., Esch, G.W., and Seed, J.R. (2001). *Parasitism: The diversity and ecology of animal parasites*. Cambridge: Cambridge University Press.

Clayton, D.H., and Moore, J. (1997). *Host–parasite evolution: General principles and avian models*. Oxford: Oxford University Press.

Combes, C. (2005). *The art of being a parasite*. (Translated by D. Simberloff). Chicago: University of Chicago Press.

Goater, T.M., Goater, C.P., and Esch, G.W. (2014). *Parasitism: The diversity and ecology of animal parasites, 2nd ed*. Cambridge: Cambridge University Press.

Loye, J.E., and Zuk, M. (Eds.) (1991). *Bird–parasite Interactions: Ecology, evolution and behaviour*. Oxford: Oxford University Press.

Marshall, A.G. (1981). *The ecology of ectoparasitic insects*. London, UK: Academic Press.

Waldbauer, G. (1998). *The birder's bug book*. Cambridge, MA: Harvard University Press.

2   LICE
It's a Beautiful Life

Clayton, D.H., Bush, S.E, and Johnson, K.P. (2016). *Coevolution of life on hosts: Integrating ecology and history*. Chicago: University of Chicago Press.

Clayton, D.H., Gregory, R.D., and Price, R.D. (1992). Comparative ecology of Neotropical bird lice (Insecta: Phthiraptera). *Journal of Animal Ecology* 61, 781–795.

Page, R.D.M. (Ed.). (2003). Tangled trees: Phylogeny, cospeciation, and coevolution. Chicago: University of Chicago Press.

3   FLEAS
The Circus in the Zoo

Krasnov, B.R. (2008). Functional and evolutionary ecology of fleas: A model for ecological parasitology. Cambridge: Cambridge University Press.

4   TOUGH TICKS

Sonenshine, D.E., and Roe, R.M. (Eds.). (2014). *Biology of ticks, 2nd ed, vols. I and II*. Oxford: Oxford University Press.

5   MITES
     Little Things Mean a Lot

Blanco, G., Tella, J.L., and Potti, J. (2001). Feather mites on birds: Costs of parasitism or conditional outcomes? *Journal of Avian Biology* 32, 271-274.

6   FLYING ZOO FLIES

Lehane, M.J. (1991). *The biology of blood-sucking insects*. London, UK: HarperCollins.
Rothstein, S.I., and Robinson, S.K. (1998). *Parasitic birds and their hosts: Studies in coevolution*. Oxford: Oxford University Press.
Webster, M.S. (1994). Interspecific brood parasitism of Montezuma oropendolas by Giant cowbirds: Parasitism or mutualism? *Condor* 96, 794-798.

7   THE WORMS THAT ATE THE BIRD

Helluy, S., and Holmes, J.C. (2005). Parasitic manipulation: Further considerations. *Behavioural Processes*, 68, 205-210.
Holmes, J.C., and Zohar, S. (1990). Pathology and host behaviour. In *Parasitism and host behaviour*. Barnard, C.J., and Behnke, J.M. (Eds), 34-63. London, UK: Taylor and Francis.
Poulin, R. (1997). Species richness of parasite assemblages: Evolution and patterns. *Annual Review of Ecology and Systematics* 28, 341-358.
Thomas, F., Adamo, S., and Moore, J. (2005). Parasitic manipulation: Where are we and where should we go? *Behavioural Processes* 68, 185-199.

8   ODDITIES IN THE FLYING ZOO

Riley, J. (1986). The biology of pentastomids. *Advances in Parasitology* 25, 45-128.

9   FLYING ZOO BEHAVIOUR

Barnard, C.J., and Behnke, J.M. (Eds). (1990). *Parasitism and host behaviour*. London, UK: Taylor and Francis.
Norris, K., and Evans, M.R. (2000). Ecological immunology: Life history trade-offs and immune defense in birds. *Behavioral Ecology* 11, 19-26.
Perez-Tris, J., Carbonell, R., and Telleria, J.L. (2002). Parasites and the blackcap's tail: Implications for the evolution of feather ornaments. *Biological Journal of the Linnean Society* 76, 481-492.
Pruett-Jones, S.G., Pruett-Jones, M.A., and Jones, H.I. (1990). Parasites and sexual selection in birds of paradise. *American Zoologist* 30, 287-298.
Walther, B.A., and Clayton, D.H. (2004). Elaborate ornaments are costly to maintain: Evidence for high maintenance handicaps. *Behavioral Ecology* 16, 89-95.

Weatherhead, P.J., Metz, K.J., Bennett, G.F., and Irwin, R.E. (1993). Parasite faunas, testosterone and secondary sexual traits in male red-winged blackbirds. *Behavioral Ecology and Sociobiology* 33, 13-23.

Zuk, M., Johnson, K., Thornhill, R., and Ligon, J.D. (1998). Parasites and male ornaments in free-ranging and captive red jungle fowl. *Behavior* 114, 232-248.

10 ENVIRONMENTAL IMPACTS
The Future of the Flying Zoo

Cleaveland, S., Hess, G.R., Dobson, A.P., Laurenson, M.K., McCallum, H.I., Roberts, M.G., and Woodroffe, R. (2001). The role of pathogens in biological conservation. In *The ecology of wildlife diseases*. Hudson, P.J., Rizzoli, A., Grenfell, B.T., Heesterbeek, H. and Dobson, A.P. (Eds), 139-150. Oxford: Oxford University Press.

Fayer, R. (2000). Global change and emerging infectious diseases. *Journal of Parasitology* 86, 1174-1181.

Friend, M., McLean, R.G., and Dein, F.J. (2001). Disease emergence in birds: Challenges for the twenty-first century. *The Auk* 118, 290-303.

Hudson, P.J., Rizzoli, A., Grenfell, B.T., Heesterbeek, H., and Dobson, A.P. (Eds). *The ecology of wildlife diseases*. Oxford: Oxford University Press.

Kutz, S.J., Hoberg, E.P., Polley, L., and Jenkins, E.J. (2005). Global warming is changing the dynamics of Arctic host-parasite systems. *Proceedings of the Royal Society of London. Series B: Biological Sciences* 272, 2571-2576.

McCallum, H., and Dobson, A. (2002). Disease, habitat fragmentation and conservation. *Proceedings of the Royal Society of London. Series B: Biological Sciences* 269, 2041-2049.

Patz, J.A., Graczyk, T.K., Geller, N., and Vittor, A.Y. (2000). Effects of environmental change on emerging parasitic diseases. *International Journal of Parasitology* 30, 1395-1405.

INDEX

Anseriformes (e.g., geese, ducks, swans), 32, 145, 147
antennae
    of fleas, 46, 48f, 50
    of lice, 20, 22f
    of mosquitoes, 101f
    ticks, lack of, 62
"anting," 39, 169
ants, 116, 201
apapane, 104
Apodidae. *See* swifts
Aptendoytes forsteri. See emperor penguin
Aptenodytes patagonicus. See king penguin
Apus melba. See alpine swift
Aquanirmus (lice), 147
Ara macao. See scarlet macaw
Aratinga holochlora. See green conure
Argas cucumerius (tick), 74
Argasidiae (soft ticks), 69
Argas persicus (tick), 73–74
Argas reflexus (tick), 3f
Arthropoda, 155–156
arthropods
    and climate change, 210
    pesticides and, 39
    ticks, 62
    tongue worms, 155–156, 155f
    as vectors of disease, 196–197
ascities, 152
Astigmata (feather mites), 89, 91–97
Athya affinis. See lesser scaup
Atkinson, Carter, 199
Atyeo, Warren, 89
Audubon, James, 209
Austrogonoides (lice), 28, 31f
avian keratin disorder, 16
avocets, 146
Aythya valisineria. See canvasback duck

bacteria
    Borrelia burgdorferi (Lyme disease), 68, 208
    endosymbiotic, 11, 20

as food for mites, 77, 94, 182
as food for nematodes, 138
relapsing fever, due to, 74
skin infections, due to, 39, 84
surface flora and preening, 181–182
Yersinia pestis (bubonic plague), 49
bacterial transmission
    by fleas, 49, 54
    by flies, 100
    by swallow bugs, 161
    by ticks, 63, 68, 73–74, 208
bald eagle, 196
Banaja, A., 157
band-tailed pigeon, 26
barn owl, 109
barn swallow, 92, 107, 172–174, 184
barred owl, 30
Bartlow, Andrew, 27
Batesian mimicry, 189
bats, 158–159
Bdellorhynchus polymorphus (mite), 90f
bed bugs, 2, 158, 159f
bees, 112, 113f
Benckman, Craig, 85
Bennett, Gordon, 107
Bethel, Bill, 123–125
bile, 131, 136, 139, 145
biodiversity
    in feather mites, 80, 91
    preservation strategies, 213t
    protection of, 165
    in relation to latitude, 200
bird bugs
    effect on nestlings, 161
    endurance of, 158, 160
    feeding behaviour of, 158
    habitats of, 158–160
    and host defense, 161
    hosts of, 158–159
    number of species of, 159
    pathology due to, 161
    physical characteristics of, 158, 159f
    population density of, 160
    as vectors of disease, 161
birds, as super worm hosts, 120

cattle egret, 70
cement, 11, 62
Centrocercus urophasianus. See sage grouse
Cepphus grylle. See guillemot
Ceratophyllidae (fleas), 52-53, 59
Ceratophyllus borealis (flea), 54t-55t
Ceratophyllus celsus (flea), 160
Ceratophyllus columbae (flea), 3f, 54t-55t
Ceratophyllus farreni (flea), 52
Ceratophyllus fringillae (flea), 54t-55t
Ceratophyllus gallinae (flea), 52-53, 54t-55t, 55-56, 185
Ceratophyllus garei (flea), 52, 54t-55t
Ceratophyllus hirundinis (flea), 52, 54t-55t
Ceratophyllus idius (flea), 48f
Ceratophyllus lunatus (flea), 46
Ceratophyllus rossittensis (flea), 54t-55t
Ceratophyllus rusticus (flea), 54t-55t
Ceratophyllus styx (flea), 52, 54t-55t
Ceratophyllus vagabundus (flea), 54t-55t
Ceratopogonidae (flies), 100
Certhidea fusca. See warbler finch
Certophyllidae (fleas), 59
Cestodes (tapeworms), 135f
Chakra, Marie Abou, 189
Charadriiformes (e.g., shorebirds), 21t, 32
Cheletomorpha lepidopterorum (mite), 79
chelicerae, 62
chelicerates, 62
chemosensor, 9, 47
chestnut-backed chickadee, 108
chicken mites, 81, 82f
chickens, 169, 184, 202, 204
chitin, 46, 134, 155
Choe, Jae, 89, 91
Christie, Phillippe, 108
Chrysococcyx basalis. See bronze-cuckoo
cigarette butts, 39
Cimex lectularius (bed bugs), 158
Cimicidae, 158, 159f
Cioniiformes (e.g., herons, ibises, storks), 21t, 32, 145

cisticolas, 189
Cizauskas, Carrie, 210
Clamator glandarius. See great spotted cuckoo
Clay, Theresa, 9, 32, 59, 114
Clayton, Dale, 26-27, 30, 34, 37
cliff sparrow, 94
cliff swallow, 159-160
climate change
    Arctic changes due to, 200
    and arthropod-associated disease, 197, 203
    contribution to ecosystem disruption, 211t
    and dinosaur extinction, 118
    due to human activity, 195
    and exposure to parasites, 197, 209-210
    impact on birds, 208, 211
    strategies to mitigate, 213t
Clitellata (leeches), 162
cloaca
    fluid, as predation defense, 191
    function of, 131, 137
    as mite microhabitat, 88
Cloacaridae (mites), 88
co-adaptation
    defined, 24
    examples of, 35, 37, 67
    rules of, 24
Coccyzus melacoryphus. See dark-billed cuckoo
Cockayne, Leonard, 164
co-evolution
    and birdsong, 178
    defined, 21
    fleas, examples of, 54, 59-60
    lice, examples of, 27-28, 31f, 33, 39-40, 44
    and mating behaviour, 184
    requirements for, 23
    rules of, 24-25, 32, 35
    in a short time frame, 199
    "sorting events" and, 29-30
    worms, examples of, 141-144, 142t

Menoponidae (lice), 21t
Merops apiaster. See European bee-eater
mesites, 152
Michaelia amplosinus (mite), 90f
Miconia robinsoniana (blackberry,
    plant), 202
Microtetrameres (nematodes), 130
midges, 100
migration, bird
    adaptation due to malaria, 103–104
    bird bug life cycle adaptation to,
        158–160
    effect of climate change on, 199–200
    effect on immunity, 194
    tick life-cycle adaptation to,
        65–66, 70
    as vector of epizoonosis, 212
migration, parasite
    flukes and, 148
    lice and, 34
    mites and, 81, 87
    nematodes and, 179
    tongue worms and, 157
Mirandornithes (grebes and
    flamingos), 146
Mironov, S.V., 97
mites
    beneficial effects of, 93–94
    evolution of, 75, 95, 115
    feeding behaviour of, 76–77, 79–81,
        88–89, 93
    and host reproductive success,
        80–81, 172
    host specificity of, 87
    as human parasites, 81
    life cycles of, 88–89
    microhabitats of, 76, 81, 86,
        88–89, 91
    nest architecture and, 79–80
    number of species of, 75, 86
    pathology due to, 81, 84, 86, 115
    phylogeny of, 77f
    physical characteristics of, 76, 89,
        90f, 91, 95
    population density of, 79, 81, 84, 93

size of, 76
taxonomy of, 75–76
transmission of, 78, 83, 85, 87
as vectors of disease, 78, 81, 196
see also feather mites, nest mites,
    nasal mites, soft tissue mites
mobility
    fleas and, 52–54
    flies and, 114
    lice and, 27
    mites and, 81, 84
model: human, environment, flying zoo
    interactions, 211f
Moller, Anders, 8, 40–41, 172–173
Molothrus ater. See brown-headed
    cowbird
Molothrus oryzivorus. See giant cowbird
Montgomerie, Robert, 186–187
mortality
    and bird malaria, 105, 199
    of chicks, 55–56, 58, 67, 105, 161
    and Newcastle disease, 202
    and scaly leg mites, 85
    see also death
mosquitoes
    avoidance behaviour of birds,
        103–104, 116–117
    effects of blood feeding by, 100
    habitats of, 117, 197–198
    as introduced species, 102, 203
    mouthparts of, 101f
    as vectors of malaria, 100, 102, 177
    as vectors of West Nile virus,
        196–197
moths (tear-drinking), 154
mottled petrel, 31f
moulting, 23, 92, 183
mountain chickadee, 108
mouthparts
    of bird bugs, 158
    of botflies, 111f
    of flies, 46, 100, 101f, 114
    of lice, 9, 18, 20, 38
    of mites, 76, 88, 90f, 93
    of moths (tear-drinking), 153f, 154

swimmer's itch, 149
synergy, 210
Szidat's Rule, 24

Tachybaptus ruficollis. See little grebe
Tachycineta bicolor. See tree swallow
Tanasia bragai (fluke), 3f
tapeworms
    co-evolution and, 143–146
    effect on other worms, 140
    feeding behaviour of, 133
    holdfast structure of, 134, 135f, 136
    intermediate hosts of, 126, 127f,
        137–138, 146
    life cycles of, 138–139
    microhabitats of, 133, 138, 140f
    pathology due to, 139
    physical characteristics of, 121f
    reproduction of, 139
    transmission of, 126, 127f
"tasty chick hypothesis," 108–109
Tatria (tapeworm), 121f, 143–144
Tatria biremis (tapeworm), 144
tawny-flanked prinia, 189
temperature, effect on
    parasite life cycle, 51, 67, 106, 138
    parasite survival, 9, 81, 91
Tengmalm's owl, 105
terns, 156
Tetrameres (roundworm), 130
Tetrameres fissipina (roundworm), 3f
Tetrameriidae (roundworm), 130
Theromyzon biannulatum (leech), 162
Theromyzon rude (leech), 162, 163f
Theromyzon tessulatum (leech), 162
thick-billed murre, 89, 91
ticks
    endurance of, 67, 70, 74
    feeding behaviour of, 62–63, 68–70,
        72, 74
    geographic distribution of, 67–68
    host defense, 74
    as human parasites, 72, 74
    life cycles of, 62–64, 69–70
    pathology due to, 68, 72, 74

physical characteristics of,
    62–64, 69–70
    population density of, 72–73
    predators of, 73–74
    reproduction of, 65–66, 70
    as vectors of disease, 63, 68, 73–74
    see also hard tick, soft tick
tinamous, 96
tits
    blowflies and, 56–57
    fleas and, 55–56
    flies and, 109
    immune response to fleas, 58
    mites and, 91, 182
    and uv reflectance, 14
tongue worms
    evolution of, 155–156
    hosts of, 156
    life cycles of, 156–157
    microhabitats of, 156–157
    phyllum debate, 154–155
    physical characteristics of, 155–
        156, 155f
    reproduction of, 157
    transmission of, 156–157
toucan, 34
toxins, 39–40, 68, 102, 142t, 174
Trabeculus (lice), 31f
transmission of disease
    and allopreening, 197
    bird bugs and, 158, 161
    fecal contamination and, 197, 202
    fleas and, 49, 54
    flies and, 100, 105, 114, 118,
        169, 177
    leeches and, 164
    lice and, 35, 36f, 172
    mites and, 79, 81, 196
    mosquitoes and, 102–103, 116, 169,
        177, 196–199
    swallow bugs and, 161
    ticks and, 63, 68, 73–74, 208
transmission of parasites, 168, 202
    ecosystem changes and, 208,
        211f, 212

lice and, 20, 27, 42–43, 205
mites and, 81, 83, 85–87, 115
nesting behaviour and, 83, 117
worms and, 123, 125, 128, 147, 149, 156
see also phoresy, hitching
tree swallow, 48*f*, 56, 78
trematodes. *See* flukes
Trichobilharzia regenti (blood fluke), 148
Trichostrongylus tenuis (nematode), 179
Trinoton (louse), 32, 147
Tripet, Frederic, 53
Trochiloecetes (lice), 8
Troglodytes aedon. See house wren
tropical fowl mites, 81, 95, 172
Trouessartia (mites), 93
trumpeter swan, 164
Trypanosoma (protozoa), 177
trypanosomes (protozoans), 114, 174
Turdus migratorius. See robin
turtle, 88
Tylodelphys podicipina (fluke), 136, 137*f*
Tyto alba. See barn owl

Uria aalgae. See common murre
Uria lomvia. See thick-billed murre
uv reflectance, 13–14, 16, 180–181, 187

vaccination, 197
Valera, Franciso, 83
Valera, Roberto, 109
vanga, 152
van Leeuwenhoek, Antoine, 60
van Ripper, Charles, 102
van Ripper, Sandra, 102
vectors. *See* transmission of disease, virus transmission, bacterial transmission
vermilion flycatcher, 203
Vestiaria coccinea. See iiwi
viruses
Buggy Creek virus, 161
equine encephalitis virus, 81, 116
flavivirus, 196

influenza virus, 212
Newcastle disease virus, 196–197, 202
West Nile virus, 114, 195–197, 203
virus transmission
by bird bugs, 161
by fleas, 49, 54
by flies, 114
by mites, 81, 196
by mosquitoes, 116, 196–197, 203
through habitat contamination, 197, 202
by ticks, 63, 73–74, 196
viviparous parasites, 87, 100, 114
vultures, 73, 155*f*, 156–157

wagtails, 189
warbler finch, 204
warblers
blood parasites and, 177
botflies and, 203-204
brood parasites and, 190, 208
and malaria, 102
mites and, 84-85, 93
song displays of, 177-178
Warner, Richard, 104
waterbirds
endemic to Galapagos Islands, 201
leeches and, 162
mites and, 93
worms and, 121*f*, 149
waterfowl
effect of climate change on, 200
and flamingos, 147
fleas and, 53
leeches and, 162-164, 163*f*
tapeworms and, 134, 139
worms and, 120, 131–132, 136
waved albatross, 68, 201
weka, 165
western grebe, 143
westland petrel, 31*f*
West Nile virus
bird illness, 196–197
global distribution of, 195–196, 203